Professional Ranks in the Roman Army of Dacia

George Cupcea

BAR International Series 2681
2014

Published in 2016 by
BAR Publishing, Oxford

BAR International Series 2681

Professional Ranks in the Roman Army of Dacia

ISBN 978 1 4073 1325 2

COVER IMAGE *Legionary standards and officers on Trajan's Column*

BAR Publishing is the trading name of British Archaeological Reports (Oxford) Ltd. British Archaeological Reports was first incorporated in 1974 to publish the BAR Series, International and British. In 1992 Hadrian Books Ltd became part of the BAR group. This volume was originally published by Archaeopress in conjunction with British Archaeological Reports (Oxford) Ltd / Hadrian Books Ltd, the Series principal publisher, in 2014. This present volume is published by BAR Publishing, 2016.

Printed in England

BAR
PUBLISHING

BAR titles are available from:

BAR Publishing
122 Banbury Rd, Oxford, OX2 7BP, UK
EMAIL info@barpublishing.com
PHONE +44 (0)1865 310431
FAX +44 (0)1865 316916
www.barpublishing.com

CONTENTS

Tables and Figures List

ACKNOWLEDGEMENTS

This work is the result of my PhD research, lead at the *Babeş-Bolyai* University in Cluj-Napoca, Romania, successfully defended in 2011, on the subject of the military ranks in Roman Dacia. The subject follows my earlier interests in Roman military history that have lead to a Bachelor thesis called *The Veterans of the Roman Army in Dacia* and to a Master thesis called *The Numerical Consistency of Roman Troops*. The idea of such a monographic study has occurred when considering, together with Prof. I. Piso, a theme for the PhD. Only approached in Romania in a shallow manner, it seemed to offer a great deal of evidence on which to work.

Many have offered their help on the way and I would like to mention a few. To Professor Ioan Piso, the coordinator of the thesis, I thank for his valuable advice and patience, for his constant support and for numerous recommendations in Romania and elsewhere. To the reviewers of the thesis, Professors R. Ardevan, C.C. Petolescu and S. Cociş I would like to thank for their opinions and advice. To my history schoolteachers, V. Gabor and T. Marton, I would like to thank for their encouragement in pursuing this fascinating domain. To my close friend and colleague Felix Marcu I thank for his perpetuous trust and appreciation.

Such an endeavor needed a good amount of time to be spent in libraries of research institutions abroad. In this purpose I owe thanks to Mme Mireille Corbier, of Paris and Prof. F. Mitthof from Vienna. Other friends from abroad also provided me with a home while in research stages; to them I am very grateful: Susanne and H. Zabehlicky, from Vienna, C.S. Sommer, from Munich and Benedicte Mămăligă, from Paris.

A special thought goes to my parents, drs. Maria and Gheorghe Cupcea, M.D.s for their care and for providing me with a solid education and to my brother, Bogdan, for his help with computer issues.

Finally, without the happiness brought into my life by my wife Adriana and my son Mihai, none of this would have been possible therefore this book is dedicated to them.

This work was supported by several grants of the Romanian National Authority for Scientific Research, CNCS-UEFISCDI, projects nos. PN-II-ID-PCE-2011-3-0096 and PN-II-RU-TE-2011-3-0273.

Translation: Cosmin Mihail Coatu.

INTRODUCTION

Historiography. Sources. Method and structure

The complex issue of the Roman army military hierarchy has been approached on many occasions, on the course of more than a century. The pioneer was without a doubt Alfred von Domaszewski who published in 1908 a monumental paper regarding the whole hierarchy of the Roman army, *Die Rangordnung des römischen Heeres*. Still, he was not the first researcher to study this aspect. During the second half of the 19th and the beginning of the 20th centuries, many works concerning military history and hierarchy were published, especially in the German academic milieu, edited by great scholars such as J. Marquardt, O. Hirschfeld, Th. Mommsen, W. Baehr, or Th. Wegeleben. From this point on Domaszewski's book was a source of inspiration and a starting point for all researchers who encountered this problem at least once. Under these circumstances, the present paper falls in one of the particular applications of the system created by the Austrian academic more than 100 years ago, namely that of the professional hierarchy of the Roman army from Dacia.[1]

Since then a series of scientists contributed to completing the results of this science, mostly in Western schools, but also in the Romanian one. Almost at the same time, G.H. Allen publishes in New York a paper similar to Domaszewski's but limited to the gathering of the multitude of posts and ranks, without any attempt to find any system of promotion or command. Also G.L. Cheesman was interested in the auxiliaries of the Roman army, making one of the first comprehensive works regarding this category of soldiers. The interwar period facilitated the publishing of monographs of reference concerning a certain category of military units such as Durry's praetorian cohorts or Parker's legions. According to the most prolific researcher from the past 50 years concerned in Roman military hierarchy, B. Dobson, researchers of Roman military hierarchy owe almost everything to E. Birley and H.-G. Pflaum.[2] Especially in the second half of the 20th century more scientists concerned themselves with this problematic, some in a general manner, others starting from particular cases. In the first category we include the major work of the two British academics D.J. Breeze and B. Dobson, the second also writing in 1967 a commented republication of Domaszewski's work. Other scientists also approached the subject of Roman military hierarchy and structure from a general point of view, among them: H. Devijver, G. Webster, Y. Le Bohec, G. Forni, C.E. Brand, W. Eck, G. Alföldy or B. Campbell. Among those that started research from particular aspects to try and present the problems of each rank comprised in this web, the most prolific were perhaps E. Birley and M.P. Speidel.[3] Next to them stand N.B. Rankov, M. Clauss, J.F. Gilliam or A.K. Goldsworthy.

In recent times another series of researchers turned on this subject, with spectacular results especially concerning the particular study of a category of troops or soldiers. We will name but a few: R.W. Davies, R.L. Dise jr., L. Keppie, J. Ott, E. Schallmayer or M.A. Speidel.

At the International Congress held at Lyon in 1994 dedicated to this matter, the proceedings being published in 1995, the discussion took in several cases a theoretical approach, very important to consider when starting such endeavors. The two great academics that contributed substantially to the study of this problem, Y. Le Bohec and B. Dobson,[4] both agree when presenting a theoretical approach of this subject. The former greatly emphasizes the theoretical presentation of the significance of each rank of the military hierarchy, a line of study that the "forefather" of this subject, Domaszewski, did not emphasize. He also mentions the need to study the problem taking into account all categories of sources that his illustrious predecessor did not have access to. The latter emphasizes the importance of the chronological factor, maybe more so than the theoretical one, in establishing the command structure and hierarchy of the Roman army, again an aspect neglected by Domaszewski.

In Romania the problem was seldom discussed, mostly in connection with other subjects or as brief explanations when encountering an officer while publishing a *corpus* of inscriptions. Initially the subject was more or less briefly discussed by V. Pârvan, C. Daicoviciu, M. Macrea. More recently new contributions on this subject came from I. Piso, M. Bărbulescu, C.C. Petolescu, Doina Benea, V. Moga, R. Ardevan or D. Isac.

With such a precedent in mind I started work on my doctoral thesis called *Military ranks in Roman Dacia.* This title does not fully explain its contents. If the superior officers of the Roman army were very recently discussed in our historiography,[5] the lower ranks were somewhat neglected.[6] It is possible that upon first seeing their great number one is discouraged or maybe because the research done by foreigners, partially including Dacia of course was considered enough that a monograph concerning the professional officers of the Roman army in Dacia was not conducted. I included in this category all professional soldiers who, while serving in the army, held one or several lower ranks, considered necessarily as professional. This does not mean that superior officers are not professionals,[7] but the transiency of their

[1] All researchers concerned with the study of Roman military hierarchy are greatly indebted to the Austrian savant, their work developing, maybe even unwillingly, from his paper. Even this present work might not have been possible without the solid base laid down at the beginning of the past century, that is why, like all other scientific papers covering this topic, it represents a tribute to Domaszewski.

[2] Dobson 1979, 204.

[3] Who managed to collect and print the most valuable studies regarding Roman military in the MAVORS series.

[4] He also presents a list of papers concerning military hierarchy, Dobson 1995, 42-46.

[5] Piso 1993 and the second volume, Piso 2013.

[6] They are mentioned in lists in Benea 1983, Moga 1985, Bărbulescu 1987, Isac 2001.

[7] Although their status as amateurs and the incompetence of military leaders during the Principate was often discussed.

command posts clearly separates them from the notion of soldiers.

Thus the inferior rank limit for our study is *miles immunis*. It is the first category of soldiers that is distinguishable by a clear attribute from the simple *miles* and opens a possibility of ranks and posts that a valuable soldier can achieve. The purpose of the Roman soldier was generally to become a centurion, to move on from executing commands to issuing them. The most important category of centurions, *primi pili* is the higher hierarchic limit of this paper. Becoming chief-centurion was the only legitimate and legal way for a soldier, a member of the *humiliores*, to climb up the social ladder, becoming a knight. Thus this rank was highly coveted at all the levels of the Roman army and competition in order to gain it, a fact that was possible only in the legions, was always fierce. Because these accomplished professionals, the *primipilares*, sometimes continue their career in the equestrian category I tried to present their role in that sector as well, because they are entirely different from their other equestrian colleagues.

Sources

Like almost all the aspects of Roman history, military history is based on at least three types of sources: literary, epigraphic and archaeological.

Literary sources

In the first category we include works of the ancient authors concerning the military aspect, what we call military *compendia,* where the works of Vegetius or Pseudo-Hyginus are of great importance. They come in addition to the more professional writings of superior officers or Roman military commanders that take the form of commentaries regarding the wars. Among them, Caesar, Josephus Flavius or Arrian offer invaluable information concerning the organization of military units and their hierarchy. Other types of literary sources are the monographs concerning a certain event or time period or the whole Roman history where the problem of military hierarchy is more or less discussed. Here we include most of the works of Roman historians such as Livy, Caesar, Appian, Suetonius, Tacitus, or of other origin, such as Polybius, Josephus Flavius and Cassius Dio. Legal sources are of great importance. The precise contributions of some great jurists such as Ulpian or Tarruntenus Paternus were brought together in collections concerning Roman law, with *Digesta* as the most important one. Last but not least even literary authors contribute to investigating military hierarchy: Juvenal, Cicero, Petronius, Apuleius or Pliny the Young. Because it was omnipresent in the Roman world, the military element could not be ignored in any major work of that period. The accuracy or skill with which the ancient authors tackle this problem differs. If some of them catalogue the Roman army in a seemingly professional manner, as is the case for Vegetius and Pseudo-Hyginus, others present it briefly or limited to a particular time period or case. Unfortunately the most complete works concerning the subject date back to a time period slightly posterior to the one we refer, thus their information cannot be used exhaustively. Because of this the research need to be critically correct, in order to avoid the involuntary traps set by the confusions from the ancient texts.[8]

The conglomerate of literary sources, some of them called military treatises - Vegetius, Pseudo-Hyginus and Vitruvius[9] - are fundamental works. Also in the literary works there is a series of substantial mentions concerning military hierarchy: Aulus Gellius,[10] Pliny the Elder, Festus[11] or Varro. Besides these one cannot neglect the related information from ancient novels, such as the episode with the matron from Ephesus in *Satyricon* or in Apuleius's *Metamorphoses.* One of Statius's poems[12] mentions the involvement of the emperor in naming officers; Juvenal[13] and Horace[14] discuss the length and difficulties of a military career as compared to the riches bestowed upon those who succeed. The letters exchanged between Pliny and Trajan are documents of paramount importance and Suetonius's imperial biographies must be completed with technical notions from Tacitus.[15]

The importance of literary works is not to be neglected when studying Roman military hierarchy. Certainly most of them do not offer precise information regarding the military hierarchy but the precise contribution of many, their desire to show people in action can offer us elements both descriptive and chronologically valuable. They certainly constitute a source that complements the epigraphic one.

Epigraphic sources

The largest body of information concerning Roman military hierarchy comes from epigraphic sources. A great number of inscriptions offered, for the scale of the whole Empire, a great deal of information, becoming mandatory for the study of Roman history. If according to Y. Le Bohec more than 10% of the total number of inscriptions refers to the military element, the richness of sources revealed to us is enormous. Epigraphic sources are also of several types. The most numerous one are inscriptions on stone, mentioning a great number of military officers in different situations. When discussing Dacia they represent almost 100% of the sources, of unparalleled richness when it comes to the information they offer. Secondly are inscriptions written on metal or wood, discharge papers - *tabulae honestae missionis* and those granting special privileges – military diplomas are especially valuable, illustrating in the case of auxiliary soldiers a different status from that of other soldiers.

---

[8] M.P. Speidel even claimed that we should ignore the literary sources, that are most of the time confusing, in order to study the problem truly detached and objectively. Speidel 1992, 29-30.

[9] Especially Book X.

[10] 1.11, 5.6, 10.25, 16.4.

[11] *De verborum significatione.*

[12] *Silvae* 5.1.94.

[13] *Sat.* 14.196-198.

[14] *Sat.* 1.1.4-5.

[15] Ducos 1995, 47-48.

Complementary, due to recent discoveries, military papers of any kind, written on *papyri* reveal a trove of information concerning all the organizational and functional aspects of the Roman army, from logistic problems to special missions and the pay of soldiers. They have proven to be of great importance when studying the Roman army and have opened a new practical perspective to this study. Put together in *corpora*,[16] or studied separately in different problems, *papyri* have proven to be one of the most valuable resources for the study of Roman military organization and hierarchy due to their particular and unique character.[17]

To conclude, it would be inconceivable not to admit the extraordinary usefulness of the computer and of the facilities that came along with it. Data bases accessible on the Internet containing Greek or Latin inscriptions, or any other kind of monument from that age, such as those created under the supervision of M. Clauss or those from Heidelberg or Salzburg have proven to be very useful to the student of Roman history, granting him the possibility of easy access to the *corpora* of inscriptions and to connect particular research to the general one.

Archaeological sources

The ancient world cannot be correctly discovered without archaeological research. It is the same for the Roman army whose history would be severely limited without archaeological research. Military archaeology has offered a great deal of information regarding this environment, implicitly the hierarchy, that proves invaluable in the process of understanding this complex mechanism. First, archaeological discoveries from military camps have sometimes helped more than the most complete literary sources in understanding the organization and number of Roman troops. Second, iconographic monuments have allowed identifying different types of ranked soldiers and their equipment. Last but not least the discoveries of military equipment have allowed for a better identification of ranks, images, nature and their functions. The information offered by the '*Benefiziarerlanze'*, the transversal crest or the staff of the *optio* was fundamental in knowing professional Roman military hierarchy. In Dacia too archaeology revealed an inexhaustible vein of resources regarding this aspect.

Method and structure

All these categories of sources as well as the precious contributions of previous or contemporary academics must form a solid base for what I have set out to do, the study of the hierarchy or the Roman army from Dacia from its professional perspective. By far the most complex aspect appears that of the hierarchy of soldiers. The wealth of information offered by epigraphic sources is the main ingredient from this point of view with Dacia having a front role as far as these characters are attested. To offer a probative example consider that in the north-Danubian province the number of attested veterans far surpasses 100 in the two Germanic provinces it barely amounts to 13. Here is thus a fertile environment for a study the likes of which I have set out to do. Unfortunately the least documented aspect in Dacia is that of promotions and careers. In this sense it is very important to understand the hierarchy of Roman military units everywhere else in the Empire, in order to recreate this apparently solid block.

The history of Roman Dacia can be understood only as part of the larger history of the Roman Empire and thus only by knowing the whole historic phenomenon. Thus I have tried to understand the meaning of the multitude of ranks and posts from the careers of the soldiers, just as they were studied in other parts of the Empire. To this goal my research went from the general towards the particular, starting from literary sources and contemporary monographic studies towards individual epigraphic sources and studies concerned with a certain category of officers or a particular phenomenon found in an inscription. It was necessary to study each category of Roman units because the connections between them are very strong, especially as far as it involves soldiers, personnel and officers as elements of the whole functional entity in the Mediterranean space.

For the benefit of systematization of my work I have chosen the classification proposed by Domaszewski more than 100 years ago. He classified Roman military ranks in several categories: soldier's ranks – *principales*, centurions and *primipili*, equestrian officers and senatorial officers. This systematization has been respected in all previous studies referring to Roman military hierarchy. Because of this and to ensure the systematization needed for a work of this size I considered applying the same structure as adequate.

The prologue of this work is an introductory one, concerned with problems connected with the historiography, different types of sources as well as presenting the method and structure. In the end I have proposed a small excursus briefly detailing the Roman army from Dacia.

The first part is concerned with presenting the military hierarchy for soldiers below the rank of centurion, entitled ***Immunes and principales.*** I have started by explaining the terminology and the general notions regarding this category of soldiers, with regards to their origin and their development during the Principate bordering also on the problem of their payment and decorations. Then I attempted a separate study for each category of military units, in the same order, legions (with a brief excursus about the troops stationed in Rome) and auxiliary troops. After a general look on the lower hierarchy for each troop category came the part referring to Dacia. To this goal I have tried to better sort the information by adding tables. When it comes to legions each situation was discussed separately because the number of evidence allowed for such a division but in the case of auxiliary troops I was forced to make a more general study because the quantity of evidence for each

[16] E. g. Fink 1971.

[17] Detailed discussion in Strobel 1995, 93-111.

unit was not conclusive. Still, towards the end of the chapter I have tried to reunite the lower hierarchy of the army from Dacia, especially in the well-established frame of *officia* of higher-ranking officials.

The second part discusses **centurions**. Methodologically speaking I have excluded from this category auxiliary centurions and decurions because their hierarchic position is clearly inferior to that of centurions from legions, a limited and individualized body of Roman professional officers. The centurionate is a difficult and complex problem. Its study must be approached from multiple perspectives but in spite of all the efforts it still hides many aspects characteristic for its nature. To this goal I have started with some general notions, containing elements about defining the rank, the origin of centurions and their access to the centurionate as well as careers of centurions and their social role. The most delicate part and probably the most difficult is, for several reasons, the one concerning centurion hierarchy. First of all the apparent lack of evidence prevents us from making any definite remarks no matter from what angle we view the problem. Second the *de facto* situation of the Roman army is still almost an unknown, the difference between theory and practice being obvious and observable when studying several types of sources. In this part I have also compelled a small excurse dedicated to the centurions of the troops in Rome who are closely connected to their colleagues from the legions, often extending their activity in legions; the opposite being also valid. To conclude, the centurions found in Dacia have been studied from as many points of view as the others from the Empire, as presented above: accessibility, career, roles, and social status. I have excluded from my paper centurions who participated in the wars for conquering Dacia but did not remain as active soldiers in the army of the province as well as those who have served in one of the legions from Dacia while they were stationed in other provinces.

A third part is dedicated to the ***primipili*** and ***primipilares.*** Although the first ones represent the highest category of centurions and the study of the two should always be conducted together,[18] I have made this distinction not absolute – I have stated many times the "embryonic" connection between these categories of officers – but to facilitate the study of the only privileged category of professional officers. To this end I started once again with a series on introductory notions, regarding the presentation of the rank or *primus pilus* and the access to the post of chief-centurion, *primipilares* and the study of their social status and their careers. In this section I also discussed the problem of the status of *primipilares* and their role in economy, society and religion. To conclude, I analyzed the *primipili* and *primipilares* found in Dacia from all the points of view mentioned above.

In the final part I attempted to reunite the main ideas of this work in the form of general conclusions where I would once again state the directions of the study and the results I reached. For a better understanding and recognition of the characters I included in the text of every chapter a series of charts and centralizing tables when needed.

Because it would be ideal to see and try to understand things, as they were not as they should be or as we would like them to be, inferior Roman military hierarchy still holds many secrets. Some of these I tried to understand and to clarify them for others by writing this paper, while still aware of my own limits and the errors that can appear at any point. Thus, even if this paper will not be a great innovation it was attempted to be very well documented, the errors or omissions present being due to the great numbers of characters I concentrated upon. The paper would like to be an application of the lower military hierarchic system from the whole Empire on the impressive quantity of sources concerning the army of Roman Dacia, to emphasize again, if needed, the impressive role of this small and ephemeral province in the history of the Roman Principate. This is not by far an exhaustive work, nor does it represent an end but rather I consider it a beginning for the profound study of each category of ranking military professionals, in particular and in general, like other already do and as it is needed also in the case of Dacia.

In the end I would like to quote from Vegetius a phrase that surprisingly is still painfully accurate, *Epitome* 2.4: *Quapropter ordinationem legionis antiquae secundum normam militaris iuris exponam. Quae descriptio si obscurior aut inpolitior videbitur, non mihi sed difficultati ipsius rei convenit inputari. Adtento itaque animo saepius relegenda sunt, ut memoria intellegentiaque valeant conprehendi.*[19]

General considerations about the Roman army from Dacia

The legions from the province of Dacia have been discussed countless times by Romanian or foreign academics, especially when connected with Trajan's Dacian wars. Some monographs were written for the three legions that were included for some time in the army of Dacia and had their garrison on the territory of the north-Danubian province *IIII Flavia*,[20] *XIII Gemina*[21] and *V Macedonica*.[22] Outside the problem has been tackled in connection to the recruitment of the legions and the settlement of veterans,[23] or in connection to Trajan's Dacian wars.[24] The most recent contributions were reunited when the papers from the congress from Lyon, in 1998, 'Les légions de Rome sous le Haut-

[18] Dobson 1979, 192-197.

[19] '*Thus I will explain the organization of the old legion, following the norms of military law. If the description seems obscure or incomplete I should not be held accountable but the difficulty of the subject. Thus, the details should be read several times, carefully so that by use of memory and intelligence they are understood.*' *Apud* Milner 1993, 34; Speidel 1992, 36.

[20] Benea 1983, reffering also to the vexillations from *VII Claudia*, that were active in Dacia.

[21] Moga 1985.

[22] Bărbulescu 1987.

[23] Mann 1983.

[24] Strobel 1984.

Empire',[25] were published; they were meant to complete and update E. Ritterling's monumental contribution.

XIII Gemina. The legion that was recruited by Caesar from Pannonia took part in both Dacian wars. It is not known for sure where it was stationed between the two wars but once the province was created it rested at Apulum, on the hilltop *Cetate*, where it also remained until the province was abandoned.[26] From here it controls all the valley of Mureş and apparently a large part of Transylvania. Vexillations from this legion take part in the construction of several auxiliary forts from Dacia, proven by inscriptions on building blocks or because of the abundance and distribution of the tile stamps produced by the legion.[27] During Trajan's reign it was not the only legion encompassed in Dacia's army being joined by *IIII Flavia*, with the garrison in Banat, at Berzobis. Between the beginning of Hadrian's reign and AD 168 it remains the only legion in the army of the three Dacian provinces, its legate being also the praetorian governor of Dacia Superior. From AD 168 onwards the legion *V Macedonica* will be brought in and from now on the provinces had one consular governor and each of the legions a praetorian legate. It participated in all the phases of the Marcomannic wars and fought for Severus in the civil war at the end of the 2nd century.[28] It remains in Dacia until the province is evacuated.

V Macedonica. It took part in both Dacian wars and was tasked with the other two legions from Moesia Inferior to organize the north-Danubian area annexed to this province.[29] To this goal as soon as the wars were over it was moved from Oescus to Troesmis where it remained until it was brought in Dacia. The exact timing of this transfer cannot be established for sure but the legion was already part of the army of Dacia under the command of M. Claudius Fronto in AD 168. In was stationed in Potaissa to support the large *limes Porolissensis.* It was from this role that it took part in the Marcomannic wars and vexillations from this legion along with others from *XIII Gemina* and the other troops from Dacia fought on the side of Severus during the civil war.[30] From now on it will take part in all events from the 3rd century alongside the rest of the army of Dacia, until the evacuation of the province.

IIII Flavia. It was created by Vespasian to succeed the ignominious *IIII Macedonica.* It was initially stationed at Burnum, in Dalmatia. At the beginning of Domitian's Dacian wars it is moved on the Danube line, at Singidunum where it will remain, although with one small break, until the 4th century. The epithet *felix* will be kept until the end of Trajan's reign. Following Trajan's first Dacian war the legion might have advanced as far as Berzobis but it is more likely until Haţeg area where it will probably station on the territory of the future *colonia Ulpia Traiana Sarmizegetusa.*[31] As soon as the province was created it will be transferred to Berzobis, in Banat, in an area of plains, in order to supervise the Sarmatian *Iazyges*, making room for the creation of the 'capital' of the new province, half-way between the legionary camps from the Trajanic period. Traces of its activity at Ulpia Traiana Sarmizegetusa are the numerous stamped tiles and the epigraphs attesting soldiers of the legion that were discovered here.[32] The legion remains in Dacia at least until the death of Trajan, when it is attested in the list of troops of Quadratus Bassus. By AD 119 it is already moved in Moesia Superior, at Singidunum.[33]

Other legions: ***VII Gemina, I Adiutrix*** and ***VII Claudia.*** The first of them, the Hispanic legion, is the legion that is the most closely connected to its province, not taking part in any campaign or expedition. Still, on an inscription from Potaissa, dated under Commodus and several tile stamps from Porolissum, dated sometime under the Severan dynasty indicate the presence of at least one vexillation in northern Dacia somewhere between the 2nd-3rd centuries.[34] The discussion concerning the role and place of the legion *I Adiutrix* in the beginning of Dacia can be approached from two angles. The first one states that because of the tile stamps belonging to the legion that was discovered at Apulum *I Adiutrix* was stationed in the same fortification as *XIII Gemina* or in a separate one, also at Apulum, at least during Trajan's reign.[35] The second one is based on literary sources[36] and archaeological research and denies the existence of a double fortress or of two fortresses proving that in fact until AD 113 only vexillations of this legion were stationed in several locations in Dacia, including *Sarmizegetusa Regia.*[37] When it comes to *VII Claudia* its active participation in the Dacian wars left us with a great deal of evidence, both tile stamps and epigraphic proofs dated during the first period of the province when it is likely that several vexillations took charge of organizing the left shore of the Danube. The legion will be absent from Dacia's history up until the time of Philippus when a vexillation will fight in the war against the *Carpi*, leaving important epigraphic marks in Dacia Inferior.[38]

Following the impressive display of troops involved in Trajan's Dacian wars only a part of the **auxiliary** troops that were used will remain in the territory of the new province. According to military diplomas that can be dated under Trajan's rule their number varies between 3-4 *alae* and 10-18 *cohortes.*[39]

---

[25] Where Piso 2000 refers to the legions from Dacia.
[26] Piso 2000, 220-221.
[27] Piso 2000, 221.
[28] Piso 2000, 222-223. For the lists of soldiers see Moga 1985, 78-129 and for the commanders Piso 1993, 210-267.
[29] Piso 2000, 212.
[30] Piso 2000, 214-215. For the lists of soldiers see Bărbulescu 1987, 63-84 and for the commanders Piso 1993, 267-281.
[31] Piso 2000, 208-209. For the lists of soldiers see Benea 1983, 203-216.
[32] On this see also Cupcea 2013.
[33] Piso 2000, 210-213. Also, at the same time a vexillation of this legion was stationed along with other three at *Sarmizegetusa Regia.*
[34] Piso 2000, 220.
[35] Mann 1983, 39-40, Opreanu 1998, 121-134.
[36] Suetonius, *Dom.* 7.3: *geminari castra prohibuit...*
[37] Piso 2000, 205-206.
[38] Piso 2000, 218.
[39] Cupcea, Marcu 2006, 190, for a list of troops.

We can claim that, according to the latest research and not taking into account the evolution of auxiliary troop movement from Dacia, a total number of approximately 15 *alae* were stationed on its territory. They are joined by an approximate number of 45 cohorts and 13 *numeri*.[40]

[40] For the detailed study of their placement in the territory and their history see, Spaul 1994; Spaul 2000; Petolescu 2002 and Ţentea, Matei-Popescu 2004, 259-296.

Part I. *IMMUNES* AND *PRINCIPALES*

I.1. General facts. Terminology

The two categories of soldiers/small officers were not clearly defined especially when it comes to the difference and limits between them. In fact they are soldiers and not officers but they function according to a very complex internal hierarchy that includes positions and ranks in accordance to the tactical and administrative needs of the unit.

It is known that all soldiers and petty officers up until the rank of centurion were considered *milites caligati* or those that serve *in caliga*. What is not known is whether centurions were included in this category, this being important for their inclusion in either the large corps of soldiers or in that of the officers. According to the few pieces of information available from literary sources, such as Seneca and Josephus Flavius, and few epigraphs it can be said that the only difference worth noting from the point of view of the Roman army is the one between *militia caligata* and *militia equestris*.[41] Also when it comes to *caligati* the Roman army makes only one distinction: between centurion and soldier.

During the Republic the mentioning of status or different ranks and posts between the *miles* and *centurio* are very few and uncertain. Polybius mentions for the half of the 2nd century BC the *οὐρόαγος*, transliterated in Latin as *optio*, the deputy of the centurion or of the decurion[42] and the standard bearers. It is certain that in the period between Marius's reform and the middle of the 1st century BC the army experienced a multitude of posts below the rank of centurion but the literary sources do not mention them at all, maybe with the exception of Caesar mentioning standard bearers. Also epigraphy is completely nonexistent from this point of view.[43]

The fact that *aquilifer* and *signifer* appear in the writings of Caesar is connected to the importance attributed by Marius to standards.[44] Still these mentions do not allow us to understand the position of these petty officers in the hierarchy of the legion. It is possible that the *aquilifer*, the most important flag bearer in the legion, was ranked only below the *primus pilus* and the *signiferi*, those that carry the flag for the whole maniple were ranked just below their centurion *prior*. Also the mentions of the payment received by soldiers do not help from this point of view.[45] When it comes to detailed descriptions of these petty officers, Caesar only mentions the similarity between the attitudes of the *aquiliferi* in battle with that of the centurions both of them being capable of the same exploits.[46] Otherwise it seems the *optiones* from Polybius's time are no longer mentioned and the musicians, although mentioned, cannot be ranked in a hierarchy.[47]

In the 1st century AD a *signifer* suggests his rank just by illustrating a *signum* on his funerary monument without making any other mention besides that of *miles*; a *bucinator* does the same. It is possible there was no need to mention the rank as long as it was illustrated. At the beginning of the Principate there was no difference between *immunes* and *principales*, the notions being almost absent, the only advantage a petty officer had at this time was an exception from *munera* as opposed to increased salary. Vegetius[48] tells us there was a difference but it referred to the food rations and not payment, the word *duplicarii* referring to those who receive double food rations that they can turn to money.[49] During the Augustan age the *principales* did not share any common traits with the officers nor had they any authority over the soldiers. This state of facts is confirmed by Tacitus[50] and Suetonius;[51] the former[52] even considers the *tesserarius* and *optio* as being just *manipulares*, simple soldiers and the carrying of the *signum* an honor for which the soldier need not be paid. Epigraphic lists of soldiers do not mention any other ranks besides that of centurion.[53]

The situation changes in the 2nd century when these *principales* who are superior to simple soldiers and who lead them, receive the right for physical punishment, just like the centurions; at the same time however they must take care of the soldiers under their command. Vegetius was inspired by Paternus when making his list, meaning these changes took place sometime under Trajan-Hadrian, but he does not expands on the hierarchy. Vegetius[54] notes that the *beneficiarii* are promoted – *promovere*, indicating a clearly different status, a change that cannot be observed with the *beneficiarii* from the 1st century. The first mention of *principales* is epigraphic – CIL VI 221 dated AD 113 and in several other subsequent inscriptions a lot of posts are considered *principales* but starting with AD 155 also the *immunes* are distinguished.[55] One notices a confusion of the two terms in the military environment but not in the juridical one. Following this supposed reform by Trajan the *immunes* are not differentiated by rank from the *milites*, they only have a different status. Although the differentiated *principales* are mentioned under Trajan they become official only from Hadrian onwards.[56]

[41] Gilliam 1946, *passim*.
[42] Polybios 6.24.2-6; 6.25.1.
[43] Harmand 1967, 345.
[44] *BG* 4.25.3, 5.37.5, 2.25.1.
[45] Harmand 1967, 346.
[46] *BG* 4.25.3-4.
[47] Harmand 1967, 347-348.
[48] *Epitoma* 1.13.
[49] Sander 1959, *passim*.
[50] *Ann.* 1.17, 20, 31, 32, 35.
[51] *Aug.* 49.
[52] *Hist.* 1.25.
[53] CIL VI 200 – AD 70 and CIL VI 228. Sander 1954, 87-90.
[54] *Epitoma* 2.7.
[55] CIL III 7449.
[56] Sander 1954, 91-93; Campbell 1994, 28.

The simple soldiers are reffered to as *munifices, gregales* or *caligati.* The notion opposed to *munus* is *officium* because the second refers to special service in a special post, following the example *omnibus officiis ex caliga.* In the 2nd century the well-known *beneficiarii* are included among the *principales*, thus the difference between the *immunes, principales* and *officiales* becomes even clearer. Careers are mentioned in inscriptions and centurions write down their last rank *ex caliga*, most often a rank included in the category of the beneficiaries (including *cornicularius*). The first such careers appear under Trajan and thus a body of petty officers that are ranked below the centurion and are considered *ex caliga* is created.[57]

This fact is supported by the many discoveries of the type *miles leg. ... aquilifer...* Also their inclusion in the ranks of soldiers is proven by the fact that the order in which they are written down in lists is based on the duration of their service and not their rank.[58] To conclude, if Vegetius still refers to *milites principales* it means that their status did not change for a long time. In the meantime the succession expressed by *miles leg. ... aquilifer...* can also illustrate the promotion from *miles gregarius* to *principalis*, just like those that occupy all three tactical ranks are called *omnibus officiis in/ex caliga.* The only rank that does not tend to make this association is *cornicularius*, albeit very rarely.[59]

In spite of the status change generated by the reform from the time of Trajan-Hadrian, *principales* receive the complete rights of superiority and authority only from Severus. He grants them the golden ring and the right to physically punish the soldiers but only in relation to minor problems. Thus it is only now that they receive the symbol of authority and are clearly differentiated from the rest of the soldiers.[60]

Another difference that must be made is between rank and function. From this point of view an inferior rank might appear before one that is considered superior because it implies an appointment of greater importance. A certain promotion order can be observed, from the *immunes*, passing through the tactical ranks, towards the categories of beneficiaries or the *officiales.*[61] There are several paths to becoming centurion. 1. By holding a rank of *principalis* in the cohorts stationed in Rome. 2. By holding important functions as *officialis.* 3. By moving from a command rank from the auxiliary troops. 4. By holding the prestigious ranks of *optio ad spem ordinis* or *aquilifer.* One cannot clearly differentiate between the posts of *officiales*, with only the *cornicularius consularis/praefecti praetorio* and *evocatus* standing out, being able to directly become centurions in the first cohort. This promotion system is altered by Severus who allows for promotion from the tactical ranks directly to centurion, probably because a centurion who had served for 25 years was no longer efficient enough on the battlefield.[62]

We must assume that every soldier sought promotion and his goal was to become centurion. The posts and ranks under centurion keep evolving during the 2nd and 3rd centuries. Apparently in the 1st century and at the end of the Republic all soldiers below the centurion had the same rank. If this was the situation it is probable that there were soldiers who were given tasks and soldiers who were not. Thus the difference is between people available for general tasks and those that only did special assignments, a fact that may or may not mean a promotion, although it is certain the man was still *miles. Vacatio munerum* was obviously a privilege in itself and its cancelation a form of punishment but the only difference compared to other soldiers is this exemption and not a different rank. The term enters usage in the beginning of the 2nd century although we have evidence of its use in the 1st century also.[63]

Several ancient authors offer information about this category. Tarruntenus Paternus, praetorian prefect under Commodus, whose writings survived through the *Digest* offers one of the most explicit lists and definitions of the *immunes.*[64] It is thus quite clear that the soldiers who do not share the work conditions of their comrades enjoy some form of immunity, useful only for performing that type of task. Tarruntenus Paternus' list can be considered complete at least for the end of the 2nd century.[65]

Unfortunately Vegetius[66] who represents the second important literary source on this subject confuses the two terms.[67] These two phrases from Vegetius make even Domaszewski to hesitate when approaching this subject, stating that *principales* are the soldiers who do not perform great labor - *munera.*[68] The confusion is also due to Vegetius including all the posts he knew among the *principales.*[69] It is for this reason that in the part of the

[57] Sander 1954, 93-96.

[58] CIL VIII 2568, 2569, 2567.

[59] Sander 1954, 96-98.

[60] Sander 1954, 98.

[61] Tactical ranks are the gateway towards becoming a centurion, practically a stepping-stone.

[62] Sander 1954, 99-103.

[63] Watson 1969, 75-77.

[64] 50.6.7: *Quibusdam aliquam vacationem munerum graviorum condicio tribuit, ut sunt mensores, optio valetudinarii, medici, capsarii, et artifices et qui fossam faciunt, veterinarii, architectus, gubernatores, naupegi, ballistarii, specularii, fabri, sagittarii, aerarii, bucularum structores, carpentarii, scandularii, gladiatores, aquilices, tubarii, cornuarii, arcuarii, plumbarii, ferrarii, lapidarii, et hi qui calcem cocunt, et qui silvam infindunt, qui carbonem caedunt ac torrent. In eodem numero haberi solent lani, venatores, victimarii, et optio fabricae, et qui aegris praesto sunt, librarii quoque qui docere possint, et horreorum librarii, et librarii depositorum, et librarii caducorum, et adiutores corniculariorum, et stratores, et polliones, et custodes armorum, et praeco, et bucinator. Hi igitur omnes inter immunes habentur.*

[65] Watson 1969, 76.

[66] *Epitoma* 2.7: *Hi sunt milites principales, qui privilegiis muniuntur. Reliqui munifices appellantur, quia munera facere coguntur.*

[67] The confusion lies in that the list is attributed by one author to the *immunes* and by the other to the *principales.* Sander 1954, 95.

[68] Domaszewski 1908, 2. Correction by B. Dobson, Domaszewski-Dobson 1967, vi.

[69] *Epitoma* 2.7: *Aquiliferi qui aquilam portant. Imaginarii qui imperatoris imagines ferunt. Optiones ab adoptando appellati, quod antecedentibus aegritudine praepeditis hi tamquam adoptati eorum atque vicarii solent universa curare. Signiferi qui signa portant, quos*

paper dedicated to the hierarchy of soldiers Domaszewski will include them all among the *principales*, this being also the title of this first part.

After taking a second look we can notice that the confusion Vegetius made can be explained. A soldier who knew how to write and read had all the chances to become a basic secretary, *librarius*. Many were directly recruited on this post. An example is that of Iulius Apollinaris who, in his correspondence, discusses with his father how he became *librarius*, exempted from the work his colleagues did and thus *immunis*. He asked the prefect of Egypt, Claudius Severus, to make him his secretary but because there were no posts available he sent him to the *cornicularius* to redistribute him as *librarius legionis*. In a second letter written to his mother he refers to himself as *principalis*. Vegetius says that *principales* are protected by their privileges. Thus we have confusion between the two categories in the sources although in practice it is quite clear that the *principales* are the petty officers that differ because of rank and payment from the simple soldiers. Legally there was a clear difference but the soldier considered any advantage gained as a promotion. The term began to be used in the beginning of the second century although it was quite clear that the ranks existed long before then.[70]

On the other hand Domaszewski noticed a classification in the *principales* that will prove vital for the study of the hierarchy of soldiers from that moment onwards. It refers to the grouping of three posts, present at the level of the *centuria*, as tactical ranks. He discusses these three ranks, *omnia officia in caliga – tesserarius, signifer, optio*,[71] and follows their origin until the time of Polybius. He also mentions that promotion to a post of superior *principalis* was possible only after holding at least one of these tactical ranks. Also the apparent confusion from the previous pages becomes clear with the Austrian author stating clearly that all posts below these tactical posts are not to be considered in the *principales* category during the Empire but rather in the *immunes'* category, quoting also Paternus' definition.[72]

Domaszewski also considers the ranks above the tactical ones as soldiers working in the commanders' staffs. They are very competent soldiers, attached to superior officers who despite the differences in status or tasks are generally known as *beneficiarii*.[73] According to the definitions given by Festus[74] and Vegetius[75] these soldiers were exempted from *munera* by high-ranking officers who could give them such a privilege. Thus the beneficiaries of the tribunes were *immunes* since ancient times. When speaking about cavalry Pseudo-Hyginus[76] uses the general term *principales* for decurions, *duplicarii* and *sesquiplicarii* while the rest are named, by contrast, *gregales*. In the time of the Empire commanders of provincial armies and military prefects have their own *principales* and their names adapt to the high-ranking officers they serve (as *beneficiarii*). The sum of the *principales*, beneficiaries and *immunes* that serve a superior officer forms his *officium* and in several cases are referred to as *officiales*.[77]

Typical for the manner in which this problem is tackled is the phrase Domaszewski uses at the end of this introductory part. 'The apparent confusion can be settled only by separately analyzing the situation for each type of troops.'[78]

The Roman soldiers are specialized in so many domains that it is difficult to define the simple soldier. The specialization of auxiliary troops is in accordance with the defining characteristic of that troop and we can see that as far as legions are involved there is a specialization of the armaments up until the 3rd century, all this being valid only for soldiers, not officers. For a *pedes* to become an *eques* would be considered a promotion; obviously so would be considered a pay rise and a transfer into a higher-ranking troop. Most soldiers are generally referred to as *miles, gregarius, pedes, munifex, eques* or *gregalis* but they are not the true specialists of the army. Those exempted from chores, the *immunes*, were the real specialists. Even though the term was probably unused in the 1st century it is almost certain that the principle existed.[79]

Generally speaking most *immunes* refer to technical or administrative functions. Soldiers with a special training or specialization were asked upon to serve their troop or their officer in making and administrating official documents, for supplies or payment or for constructing buildings or other duties related to *fabrica*.[80] Also they were available for the emperor or their superior officers to delegate them to overseeing or even effective

nunc draconarios vocant. Tesserarii qui tesseram per contubernia militum nuntiant; tessera autem dicitur praeceptum ducis, quo uel ad aliquod opus uel ad bellum movetur exercitus. Campigeni, hoc est antesignani, ideo sic nominati, quia eorum opera atque virtute exercitii genus crescit in campo. Metatores qui praecedentes locum eligunt castris. Beneficiarii ab eo appellati, quod promoventur beneficio tribunorum. Librarii ab eo, quod in libros referunt rationes ad milites pertinentes. Tubicines cornicines et bucinatores qui tuba vel aere curvo vel bucina committere proelium solent. Armaturae duplares qui binas consecuntur annonas, simplares qui singulas. Mensores qui in castris ad podismum demetiuntur loca, in quibus tentoria milites figant, vel hospitia in civitatibus praestant. Torquati duplares, torquati simplares; torques aureus solidus virtutis praemium fuit, quem qui mervisset praeter laudem interdum duplas consequebatur annonas. Duplares, sesquiplicares: duplares duas, sesquiplicares unam semis consequebantur annonam. Candidati duplares, candidati simplares. Hi sunt milites principales, qui privilegiis muniuntur. Reliqui munifices appellantur, quia munera facere coguntur.

[70] Watson 1969, 77-78.

[71] Domaszewski 1908, 2-3, later on, two more posts of military *principales*, *imaginifer* and *aquilifer* are created (Domaszewski 1908, 4).

[72] Domaszewski 1908, 2-3.

[73] Of these the rank of *beneficiarius tribuni* has a different status.

[74] *Verb. signific.* 33: *beneficiarii dicebantur, qui vacabant muneribus beneficio.*

[75] *Epitoma* 2.7: *beneficiarii ab eo appellati, quod promovebantur beneficio tribuni.*

[76] *De munit. castr.* 16.

[77] Domaszewski 1908, 4-5: CIL VIII 9763, 18042, 17639, III 14387, 4311, 14214, V 8275 and XI 4182.

[78] Domaszewski 1908, 5.

[79] Speidel MA 2001, 50-52.

[80] A fresh registry of *fabri* has been recongnized recently in some Tabulae Vindolanenses, TVind IV-1, 862.

construction of public buildings. To this end the legionaries from the legion *III Augusta* built the Trajanic city of Thamugadi.[81]

The rank of these *immunes* was not superior to that of normal soldiers but it still implied a privilege. Exemption from chores is a privilege that was apparently awarded to many soldiers but epigraphic proofs indicate just a 10% of all legionnaries.[82] *Immunes* were not completely exempted from physical activities but it is highly probable that they enjoyed a special status in the work they took part in. The above-mentioned example of Iulius Apollinaris is relevant.[83] In AD 107 he writes to his father, a legionary *signifer*, that he must not work in the quarry or do any other chores all day long because he is the *librarius* of the legion, hoping to promote in the governor's *staff*. A year later he is promoted at Bostra and writes approximately the same to his mother, with the sole exception that he refers to himself as *principalis*.[84] In a third letter, ten years later he says he is *frumentarius leg. III Cyrenaicae*.[85] The fact that in the second letter he uses the term *principalis* to refer to an *immunis* type post indicates that the difference between these terms was not yet well established.[86]

These pieces of evidence, alongside other similar ones, prove that alphabetization was important criterion for promotion, even when it came to post of a more military nature, such as *signifer*. It is more likely to be so in the *auxilia* and the fleets because we can see that the level of alphabetization was so high in the legions it did not represent a specialization. Thus administrative experience alongside military skill was a very important criterion when promotions were considered. There are many other categories of specialists that exist in the army beyond those we have exemplified here. The Roman army tried at least in theory to offer each and every recruit the possibility to climb up the social ladder by promotion. It is quite obvious that the dream of every soldier was to become a centurion, just as it was the dream of every centurion to become *primus pilus*. For some of them such a dream could become reality, careers beginning as low as *miles* and going all the way to *primus pilus* or even higher being not necessarily exceptions. Still for most of the soldiers even obtaining *immunitas* was an achievement. Even though they were not trained in their skills before enlistment they could gain such knowledge during military service; evidence for this comes from the multitude of the *discentes* that are attested.[87]

Apollinaris' example confirms that jobs in the personnel serving the officers or in other groups of specialists were limited, being thus necessary to wait for vacancies. Recommendation was also very important. It is possible that Apollinaris benefitted from the favor of his father's colleagues. Although the number of attested military careers is rather small some tendencies can be observed. It is probable that as far as most specialists in the army were interested knowledge served for nothing more than gaining *immunitas*, the recognition of their merits taking place in their military environment rather than the civilian one. This might be the reason why most of them chose to remain close to their former units even after they had been discharged.[88]

[81] CIL VIII 17842, 17843; in the same manner the legionaries from the *IIII Flavia* built Ulpia Traiana Sarmizegetusa.
[82] CIL VIII 7449.
[83] P. *Mich*. 466.
[84] P. *Mich*. 465.
[85] P. *Mich*. 562.
[86] Speidel MA 2001, 55-56.
[87] Speidel MA 2001, 56-58.
[88] Speidel MA 2001, 59-61.

I.2. The legion

I.2.1. Excurse. The troops in Rome

The praetorian cohorts were highly privileged, high enough that an *ex caliga* praetorian was considered superior to a legionary *principalis* or to many officials from the staff of the procurators.[89] In order to join the ranks of the praetorian cohorts a soldier had to pass through the ranks of the *vigiles* and/or urban cohorts, where he would hold any ranking posts.[90] Besides the simple *milites praetoriani* there were many other posts, the people serving here being used in the staff of superior officers or even in that of the emperor himself.[91]

Officium praefectorum praetorio. Cornicularius praefecti is the highest-ranking position in this structure. Instead of *commentariensis* the praetorian cohorts seem to have the *a commentariis*, with the variant *custodiarum.*[92] He is followed by the *beneficiarius praefecti, a questionibus praefecti,*[93] *singularis praefecti.* Evidence for other types of petty officers that were part of *officium praefecti praetorio* appeared after Domaszewski. We refer to the *scriniarii*, a type of secretaries typical of the 3rd century. *Scriniarius praefectorum praetorio*, alongside the *primoscrinius castrorum praetorianorum*, can be promoted to *ostiarius praefectorum praetorio.* They are assisted by an *adiutor comentariorum at scrinia praefectorum*, a former soldier in the urban cohorts.[94] *Ostiarius praefectorum praetorio* advances to *canalicularius* or *laterculensis* (a type of *beneficiarius*) who moves directly to centurion in the urban cohorts.[95] Also the *exceptor* is differentiated from the other secretaries by the swiftness with which he can write down messages or documents that are to be transmitted, thus justifying also his presence in the *stationes.*[96]

In Dacia there is only one character attested to have had any military rank or post in the troops from Rome. He is a soldier of *cohors I Praetoria, scriniarius praefecti praetorio* and decurion of the *colonia.* He is part of a famous military family in Sarmizegetusa. Alongside his brothers, Valerianus, beneficiary of the consular, Carus, *frumentarius* and his sisters Valeria and Carissima he honors his father, C. Iulius Valerius, a veteran of the legion *XIII Gemina Severiana*, a former *beneficiarius consularis*, decurion and *IIvir* of the *colonia.* His three sons, who are as famous as he is, are honored by their exceptional acceptance into the decurion order while still serving in the military, without being allowed to exercise their prerogatives.[97] The three are exceptional soldiers, two legionaries and one praetorian, the latter being a petty officer working as a secretary in the *officium praefecti praetorio.*

Name	Unit	Post/Rank	Source	Location
C. Iulius Fronto	*coh I praetoria*	*scriniarius praefecti praetorio*	AE 1933,248 =IDR III/2 113	Sarmizegetusa

Officium tribuni cohortis is headed by the *cornicularius tribuni*,[98] followed by the *beneficiarius tribuni.* Later the difference between the tactical ranks of cavalrymen (*vexillarius*[99] and *optio equitum*) and those of the infantry (*signifer, optio, tesserarius*)[100] will be made.

The category of the *immunes*, most of them technicians, starts at this point. The highest-ranking ones are the musicians, *tubicen, cornicen, bucinator.*[101]
Apparently in the case of urban and praetorian cohorts *custodes armorum* are replaced by *fisci curatores*, who account for the weapons and are attested only in this type of units.[102] *Optio carceris, optio valetudinarii* and *victimarius* fall in between technical and administrative posts, forming a special group that includes the *architectus* (most likely a function rather a rank, the mention *ordinatus* proving his military status, the fact that he is enlisted in the troop as well as a special status, that of a technical officer, being the equivalent to the rank of centurion),[103] *librator*, also a construction technician, similar or equivalent to *mensor*,[104] one for each cohort, with well-defined tasks in drawing maps[105] (*discens mensorem* is the apprentice of a *mensor*[106]), *custos vivarii, venator immunis.*[107] The army used some of these technicians quite efficiently; especially those in the praetorian cohorts could have access to successful careers.[108] Instructors (*campidoctor, exercitator armaturam*) and medical personnel (*medicus clinicus, veterinarius*[109] and *chirurgus*[110]) are included in the same category.

89 Domaszewski 1908, 20.
90 Durry 1968, 97.
91 Passerini 1939, 74. Recruited only from Italy and the most Romanized neighboring provinces. Šašel 1972.
92 Stauner 2004, 179.
93 Domaszewski 1908, 21.
94 Stauner 2004, 181 considers him part of the *officium praefecti Urbi.*
95 Domaszewski-Dobson 1967, x.
96 Two *exceptores* under the surveillance of a *beneficiarius* are stationed at Vazaivi (CIL VIII 17634). Stauner 2004, 141-142.

97 Ardevan 1987, 121; Ardevan 1991, 169.
98 Domaszewski 1908, 22: CIL II 2610, III 7334, VI 2440, XI 5646.
99 For the praetorian cohorts only *vexillarii* and *signiferi* are attested the former in a rather small proportion (3:33). Alexandrescu 2010, 169.
100 Domaszewski 1908, 23-24. *Equites praetoriani* are enlisted in the *centuriae* just as the legionary ones but their exact number is unknown. Passerini 1939, 69-70.
101 Domaszewski 1908, 24; Durry 1968, 100-101; Passerini 1939, 75.
102 Tacitus, *Hist.* 1.38, 2.67; Speidel 1992b, 132, 134-135.
103 Domaszewski 1908, 24-25; Stoll 2001, especially 302-207, 367.
104 Arnaud 1995, 251.
105 Sherk 1974, 549-550. He equals him to *caelator*, because he inscribes the measured map on bronze sheets.
106 There is a series of *mensores* from the praetorian cohorts, *qui in officio limitabant*, who activate under Antoninus Pius. They establish topographic boundaries under the authority of a centurion or tribune. Arnaud 1995, 253.
107 Domaszewski 1908, 24-26; Durry 1968, 98. Also attested in the legions, Epplett 2001, 212, 219-220.
108 Speidel MA 2001, 60.
109 Durry 1968, 99 says it is the absolute proof for the mixed character of the cohorts.

The situation from the *urbanae* and *vigiles* cohorts was more or less the same, the cohorts from Rome functioning as a whole from this point of view.[111] *Equites singulares Augusti* have the lower hierarchy very similar to that of auxiliary cavalry units,[112] and *castra peregrina* seems to be a training stage for the information petty officers used by the emperor.[113]

[110] Domaszewski 1908, 26, completed by Domaszewski-Dobson 1967, x. For his special status, that of a technical officer, Stoll 2004.
[111] Details in Domaszewski 1908; Freis 1962; Clauss 1973; Breeze 1974; Bérard 1995; Sablayrolles 1996; Stauner 2004.
[112] Speidel 1994.
[113] Hirschfeld 1891; Hirschfeld 1893; Baillie Reynolds 1923; Sinnigen 1961; Clauss 1973; Mann 1988; Rankov 1990.

I.2.2. The lower hierarchy of the legion

It is considered that there were approximately 600 *immunes* and 500 *principales* in a legion. The general belief is that every *centuria*, thus every centurion, had under his direct command a *signifer*, *optio*, *tesserarius*, *cornicen* and *custos armorum*, the last two are *immunes*, *signifer* and *optio* are *duplicarii* and *tesserarius* is a *sesquiplicarius*.[114]

Officium consularis. During the Republican period the magistrates elected members in their staff from civilians on posts of *apparitores, scriba, viatores* or *praecones*. In the provinces the governors were helped in administration affairs by a *quaestor* with his *scriba*, a *consilium* of his friends and *apparitores*, probably a total of 20-30 men, organized *ad-hoc* and made up of amateurs. Moving on to the Principate the administration of provinces is changed, initially with elements from the *familia Caesaris* that become specialized bureaus, especially in imperial provinces. In the beginning Pompey, after *lex Gabinia*, then Caesar, the triumvirs and in the end Augustus preferred to hire *legati pro praetore* as administrators of their provinces. These administrators could not recruit *apparitores*, with the possible exception of lictors, because they were not officially sent by the Senate but received their prerogatives from the magistrate (the proconsul). Still, these *legati* had under their command plenty of soldiers in the armies of their provinces, a recruitment pool for people skilled in administrating the army and the whole province when *apparitores* were missing. Only citizens were chosen, a direct influence on the later practice of choosing just legionaries. These practices appeared around 50 BC; until the half of the 1st century AD the system of using soldiers in the administration of more than three quarters of the provinces of the Empire was completed.[115]

The Roman army developed an entire bureaucratic apparatus, made up of several types of *officia*, one for each tribune and prefect for the auxiliary troops, one for each tribune, legionary legate, *praefectus castrorum* and one for the governor. The simplest ones include a *cornicularius* and a few beneficiaries but the one of the governor includes people trained in different tasks, sometimes revealed by their names. The higher is the rank of the superior officer, the higher the rank of his *officiales*. By the 3rd century we have epigraphic evidence for centurions heading provincial *officia* – *principes praetorii*, probably helped by a deputy – *optio praetorii* and by an assistant – *adiutor principis praetorii*.[116]

The provincial governor's staff was a more advanced version of the *officium* that was established for all troops by Augustus. The titles carried by the *officiales* originate in the Republican hierarchy.[117] It is not known for certain how the *officium consularis* was institutionalized but it

[114] Junkelmann 1986, 109.
[115] Rankov 1999, 15-17.
[116] Austin, Rankov 1995, 150.
[117] Domaszewski 1908, 73-74; Austin, Rankov 1995, 149.

seems to have formed around a category of soldiers named *beneficiarii*. Initially they were but simple soldiers exempted from chores because of a *beneficium* offered to them by a superior officer.[118] Once the army was established and maintained by Augustus, *officia* for the commanders of the troops started appearing, following the model of the governors but on a different scale. *Speculatores*, the 'spies', outrank *beneficiarii*, and *cornicularii* head these staffs. Although it is equally possible that the governors had also centurions in their proximity and the titles *princeps praetorii* and *optio praetorii* would seem to suggest they controlled the whole *praetorium* of the governor and not necessarily just the *officium*. These three ranks are the highest ones found in the bureau of the governor and they probably received double payment. The *frumentarii* are a little lower, their origin unclear connected to the development of their role during the Principate, when, apparently starting from Trajan they will create some sort of secret service of the Empire and be paid as *sesquiplicarii*. It is possible that provincial *officia* were established around these four ranks. Although they are all attested under the Flavians previous clues suggest they might have been established in this way before. Up to the 2nd century each *officialis* refers to himself with his title followed by the name of the governor; from the Antonine period onwards it is replaced with the general term *consularis*.[119]

We do not know the complete structure for these *officia* during the 1st century because several positions are not mentioned during the Republic. For example *commentariensis* is second to the *cornicularius* only sometime in the 1st century as responsible for the legal comments of the governor. Apparently they originate in the *officia* of governing procurators, where they were initially slaves from the *familia Caesaris – a commentariis*, later to be replaced by soldiers. Other positions that appeared are all *immunes*: *adiutores, exceptores, notarii, librarii, interpretes, immunes consularis, quaestionarii, haruspices* and other inferior adjutants. Guards are made up of *stratores* – legionaries and *singulares* – auxiliaries, probably evolved into *protectores*.[120]

There were always rules regarding the size of the *officium consularis*. Tacitus[121] says that in AD 39, when Caligula gave to the legion under the proconsul of Africa its own legate, who became also governor of Numidia, the two had to equally split their staff.[122] Each provincial *officium* had probably 2-3 *cornicularii* who head these bureaus when the *princeps praetorii* is missing or even alongside with him (in some inscriptions the governor's bureau is referred to as *officium corniculariorum consularis*). Next come the *commentarienses*, then *speculatores*, *beneficiarii consularis, frumentarii, quaestionarii*, and petty officers with technical duties - *interpretes*, *haruspex*, or secretaries - *librarii, exacti, exceptores*. The higher ranks *cornicularius, speculator, beneficiarius* have more general duties, greater responsibilities and the lower ones perform specialized duties.[123]

Starting from this scheme one can try to recreate the higher-end posts in the *officium* of the governor of a province with one legion: two *cornicularii*, two *commentarienses*, ten *speculatores*, around sixty *beneficiarii*, and ten *quaestores*. These numbers indicate a ratio of one *speculator* and one *quaestor* per legionary cohort and one *beneficiarius* per century, a seemingly logical fact given the Roman army. If this reenactment is correct then we must add to this group of 80 or so individuals one *princeps praetorii*, his *adiutores*, the *frumentarii* and the multitude of secretaries. The case of the *frumentarii* is a special one, they also belong to the *castra peregrina*, that hosts about 400 military men, so about 12-13 from each legion in the Empire. The secretaries were probably quite numerous. At Lambaesis[124] 22 *librarii* who are working in the office of the *prefectus castrorum* are mentioned, alongside other 28 veteran *librarii*. The governor had even more at his disposal; his office could include over 100-150 people. Extending the situation to what we know of Aquincum and Carnuntum we can have in principle some 200-300 people in the staff of the governor of a two legion-province and 300-450 for three legions. To these we add the *stratores* who act as guards, supply the provincial army with horses and have other administrative duties, probably a total number of 200 with their own officers, plus the *singulares*.[125]

The duties of this staff were very numerous and were the same as those of the governor. The functionaries assist the governor in his most important task – the exercise of justice in the province. In this case both literary and epigraphic sources refer to *officiales* actively taking part in the act of justice: a centurion takes statements, a *princeps* judges in a property dispute, *cornicularii* supervising trials and capital executions,[126]

[118] Festus, Caesar, *BC* 1.75.2, 3.88.4.

[119] Rankov 1999, 18-20.

[120] Rankov 1999, 21-23, with a complete table.

[121] *Hist.* 4.48.

[122] Rankov 1999, 23-24.

[123] Austin, Rankov 1995, 151. The members of this staff are attested by several epigraphic sources. CIL VIII 2586 and AE 1917/18, 57, from Lambaesis, numbering two *cornicularii*, two *commentarienses*, four *speculatores*, thirty *beneficiarii consularis*, four or five *quaestionarii* and one *haruspex*. No secretary posts are mentioned and probably only half of the superior staff, because we know that the legate of the *III Augusta* legion who at the same time was the governor of Numidia shared his staff with the proconsul of Africa and the number of *speculatores* in each legion is ten. At Tarraco, the headquarters of the governor of a province with one legion, CIL II 4122 enumerates two *cornicularii*, two *commentarienses*, and ten *speculatores*. CIL III 3524, from Aquincum, numbers twenty *speculatores*, and CIL III 4452 from Carnuntum (three legions) mentions three *cornicularii*, three *commentarienses*, and thirty *speculatores*.

[124] CIL VIII 2560, 2626.

[125] Austin, Rankov 1995, 152-153. Although it is possible these calculations are too rigid and things were not so exact one thing is certain and very important: the size of the governor's staff was directly proportioned to the size of the army under his command.

[126] The leadership position of the *cornicularius* is attested by the wording *officium corniculariorum (consularis)*, from the 3rd century. Rankov 1999, 19-20, 26. Details about the status, duties and careers in Domaszewski 1908, 31; Domaszewski-Dobson 1967, xii; Clauss 1973 17; Gilliam 1976; Austin, Rankov 1995, 150; Rankov 1999, 26-27; Stauner 2004, 118-126.

commentarienses serving as jailors,[127] *speculatores* executing the capital punishment by the sword,[128] *beneficiarii consularis*,[129] *stratores*[130] and *frumentarii*[131] making arrests and assisting at executions, *quaestionarii*[132] interrogating and torturing. Among other tasks the *speculatores*, *beneficiarii*, *frumentarii*, *singulares* are messengers throughout the Empire.[133] The main function of the *officium consularis* was to administer the army. Ever since the professional army was established and later on in the army of the Principate there was an archive and a bureaucratic apparatus for each unit. In these archives all categories of documents regarding the military developments from a province are included, thus it is easy to see why the 200-300 people came in hand.[134] The secretaries from the governor's office are divided in three categories: *librarii*,[135] *exactes*, *exceptores*,[136] obviously aided by *adiutores*.[137]

The symbol of these executive agents of Roman administration in the territory is the so-called *Benefiziarlanze*, stylistically represented on several monuments belonging to *frumentarii*, *beneficiarii* or *speculatores*. In reality it probably is an insignia of that shape discovered in several provinces.[138] The success enjoyed by these provincial *officia* spreads to Italy, the officers here organizing their staffs in the same manner.

Besides the *officium consularis* each superior officer has his own staff. Most *principales* who serve in this staff are *beneficiarii* and their hierarchy depends on that of the officers they serve.[139] They exist only when the legion's legate is not at the same time the governor of the province.[140] *Tabularium legionis*, bureau and archive of the legion is headed by a *cornicularius* who alongside the *aquilifer* is the highest-ranking *principalis* position in the legion.[141] Compared to the governor's staff the legate of the legion has the *actarius*[142] but the *commentariensis* is missing. Also he has a number of simple secretaries, *librarii, cerarii*.[143] Other superior officers who have the right to form *officia* organized in a similar manner are the tribunes, *praefectus castrorum* and the centurion of the first cohort *princeps prior*.

Upon finishing the review of the personnel from these *officia* we move to the category of **military ranks**. Here the most important roles belong to the standard bearers[144] and to the adjutant of the centurion, *optio*. Epigraphic and papyrological evidence for the standard bearers of the legion are quite numerous: *aquilifer*, *imaginifer*, *vexillarius*, *signifer*. In the legions there are no flags for the cohorts, only for the maniples and/or *centuriae*. Only the horsemen, vexillations or the unit of veterans use the *vexillum* in the legion. For the *signiferi* the categories mentioned *signifer legionis*, *cohortis* and *centuriae*. *Discentes* probably are apprentices, not only in the tactical meaning of the flags, but also in the system of depositing the soldiers' money.[145]

The hierarchy of the *principales* is headed by the *aquilifer*.[146] He can promote directly to centurion, meaning his position is equivalent to that of the governor's *cornicularius*. The seniority of the position is confirmed by the promotion from *signifer* to *aquilifer*. Even Vegetius[147] places him as head of the hierarchy of the *principales*.[148] This very important position is due to the sacred role of the eagle. Still direct promotion to centurion is not a rule; the post is rather connected to the rewarding of the accomplishments of many soldiers during their military stage, especially because of the great prestige that the post implied.[149] The *imaginifer* is ranked above the tactical ranks and similar to the *vigiles* under the *beneficiarius procuratoris*. His special relation to the *aquilifer* and the privileged status come from the sacred law of the army.[150]

Optio is one of the few ranks that receive a special insignia – the cane (*hastile*). It is not a position whose responsibilities have been clearly established, the sources

[127] Domaszewski 1908, 31; Haensch 1995, 268-276; Stauner 2004, 126-128.
[128] Domaszewski 1908, 32; Clauss 1973, 59, 69-77; Rankov 1999, 26-27.
[129] Rankov 1999, 27. There are 2 recent referential works concerning the *beneficiarii*, Ott 1995 (reviewed and completed by Stoll 1997) and Nelis-Clément 2000, dealing in principle with the same problems and with the same conclusions. Regarding the evolution of the *stationes* and the activity of the beneficiaries see also Hirschfeld 1891, Zwicky 1944, Speidel 1992a, Rankov 1994, Mirković 1994, Schallmayer 1994, Isac 1994, Ott 1994, Dise 1997, Dise 1997a, Rankov 1999.
[130] Domaszewski 1908, 35-36; Speidel 1974, 543-544; Speidel 1978, 11, 25, 55, 133.
[131] Hirschfeld 1893; Domaszewski 1908, 34-35; Clauss 1973, 82-110; Mann 1988, 149-150; Austin, Rankov 1995, 137.
[132] Domaszewski 1908, 34; Domaszewski-Dobson 1967, xii; Rankov 1999, 22-23.
[133] Austin, Rankov 1995, 153. One episode is relevant for the internal hierarchy: a senior *beneficiarius* forwards a petition for the emperors Valerian and Gallienus to a *kanaliklarios* (*cornicularius*) working in the bureau of the prefect of Egypt (*P. Oxy.* 47).
[134] Rankov 1999, 25.
[135] Domaszewski 1908, 37; Stauner 2004, 132-138.
[136] Berciu, Popa 1964; Stauner 2004, 141-142.
[137] Stauner 2004, 147-148.
[138] Rankov 1999, 31-32. For variants of *Benefiziarerlanze*, Eibl 1994, Ubl 2001 and Kleiser 2003.
[139] Junkelmann 1986, 110.
[140] Junkelmann 1986, 110; Nelis-Clément 2000, 114. *Contra* Ott 1995, 26 and Stauner 2004, because of the examples of Iulius Appolinaris and Cassius Longinus, from the East.
[141] Junkelmann 1986, 110; Stauner 2004, 162-165.
[142] Domaszewski 1908, 38-39; Stauner 2004, 129-131.
[143] Domaszewski 1908, 39; Domaszewski-Dobson 1967, xiii; Stauner 2004, 139-141.
[144] On standards, their role and significance see Domaszewski 1885, Webster 1986, Alexandrescu 2005, Künzl 2008, Alexandrescu 2010.
[145] Alexandrescu 2010, 169-170.
[146] The courage of an *aquilifer* when Britannia was invaded – Caesar *BG* 4.25. Webster 1969, 134. Domaszewski 1908, 41; Speidel 1976, 143; Junkelmann 1986, 110.
[147] *Epitoma* 2.7.
[148] Domaszewski 1908, 41.
[149] Speidel 1976, 143-145; Junkelmann, 1986, 110: Though most of the examples tend to indicate that the post was held towards the end of one's military career, the person holding it rarely being promoted to centurion, there are examples of *aquiliferi* who have served for less than one year. Still it is obvious that the post had great prestige attached to it, *duplicarius,* all promotions from here onwards being almost exlcusively to centurion
[150] Domaszewski 1908, 43. They are considered *duplicarii*, they are not well attested as having careers and they were probably discharged from this post. Alexandrescu 2010, 171-172.

mentioning it as having both administrative and military duties. When placing in the hierarchy of the soldiers from a *centuria* he is immediately below the centurion, so we might say he is the highest-ranking soldier, from the military point of view. Anyway on the battlefield it is his responsibility to maintain the line of battle, from this point of view his authority being similar to that of the centurion. Outside the *centuria* he can lead small vexillations. Papyrological evidence presents them while exercising their administrative duties, most of them connected to supplying the *centuria.* The special categories of *optiones*, *optio ad balneas*, *valetudinarii*, *navaliorum*, might indicate that they direct the activity of a special installation or that they lead people with specific tasks (*optio ballistarii, plumbarii*). In civilian or military associations – *collegia*, *optio* does not appear as a military rank but as some type of president of that association, always mentioned at the top of the list. In the military hierarchy *optio* is mentioned in three positions: in the lower ranks (*optio valetudinarii, carceris*), among the tactical positions (*optio centuriae*) or in the highest rank (*optio spei*, he is placed beneath the *aquilifer*, except he is much more useful and can be promoted). *Optio spei*[151] ends up here from a position in a bureau and probably is trained in the military duties of the centurion where he will eventually end. His duties can be identical to that of the *optio centuriae*. The differences between the *optiones* can be identical to those between the centurions because they are also named after the *centuria* they originated in, especially the *optio primipili, optio principi* and *optio hastati.*[152]

Signifer,[153] *optio*,[154] *tesserarius*,[155] known collectively as **tactical ranks** share the same characteristics with those from the cohorts in Rome, especially the close connection between them. They are needed in order to promote to beneficiaries of superior officers.[156]

The list comprising ***immunes*** begins with the military musicians. They serve on the battlefield as adjutants to the standard bearers, completing their tactical role.[157] The problem of communication inside the legion comprises two coordinates: visual and auditory. Obviously the orders could not be relayed orally, military musicians exist especially for this and they draw the attention of soldiers to the standards – *signa.*[158] Military musicians had an *optio* leading their association, aided by a *princeps.*[159] It is in fact quite normal that musicians, both military and civilian would be organized in *collegia.*[160] It is possible that the legion had all three categories of musicians although this is attested only for the *III Augusta* and *V Macedonica.* Their number is uncertain, only mentioned in the lists of *collegia*, just like their belonging to *centuriae* or to a separate corps of musicians.[161] In theory it is admitted that each maniple had one *tubicen* and one *cornicen.*[162]*Bucinatores* are much less attested, maybe just one per cohort, at most 20 for the whole legion. They promote to *duplicarius* and are mistakenly called *principales.*[163] Military musicians generally kept their post until being discharged. Their possibilities for promotion are limited.

Armorum custos had to keep the weapons locked away and give them to soldiers only when they needed them. However it is unlikely that they were locked or deposited based on units but rather on categories (*arma antesignana XXX* and *postsignana XIX* - Lambaesis). It is only in Rome that the praetorians and the *urbaniciani* have the weapons locked away in a central arsenal for reasons of security;[164] on the *limes* free access to weapons was needed at all time. In consequence the *armorum custos* has to be considered a petty officer charged with fiscal and supply duties. In the case of *equites singulares Augusti armorum custos* is usually an old and experienced soldier, often someone who inherited his dead colleagues,[165] present in every *centuria.* In inscriptions[166] dating from the time of Severus 62 are mentioned but this number could indicated the doubling of the first *centuria.* On the other hand we find out from a different source[167] dating from the period of Severus Alexander – Valerianus, about only 32 such men; we can conclude that each *centuria* from the first cohort has a *custos armorum* (just like in the case of the *cornicen*) and that in the cohorts II-X there was one in every maniple.[168] This post disappears at the half of the 3rd century because from now on weapons are borrowed to every category of troops, being returned after discharge.[169]

There is a separate hierarchy for the medical staff. Vegetius[170] says that *praefectus castrorum*, the second in

[151] *Optio spei* or *ad spem ordinis* is a special category of the tactical post of *optio*, that is ranked immediately below the centurion and awaits an opening so that he can obtain it, as noticed from the promotions. Such *optiones* are delayed with *missio* a few years The inscriptions show that they are the equivalent of the *candidati*, this statement only supporting what was stated above about them awaiting promotion. Because of his high position he promotes directly to centurion just like the other three degrees above the simple *optio centuriae*. During peacetime when the *centuria* does not have a leader, promotion is direct, especially in the 3rd century when it becomes a habit. It is for this reason that Vegetius places him in his list above the tactical rank of *signifer*. Promotion can be received following the suggestion of a superior officer in whose service he is, especially in the period after Severus. Junkelmann 1986, 109. CIL III 11135. Domaszewski 1908, 41-42.

[152] Probst 2009, 1113-1115.

[153] Domaszewski 1908, 43-44; Speidel 1982, 856-857; Junkelmann 1986, 109-110; Isaac 1995, 28.

[154] Domaszewski 1908, 44; Domaszewski-Dobson 1967, 44, *contra* xiv; Breeze 1976, 127-128; Speidel 1992, 24-26;

[155] Vegetius, *Epitoma* 2.7; Webster 1969, 30-31.

[156] Domaszewski 1908, 43.

[157] Durry 1968, 100-101.

[158] Caesar, *BC* 3.46, 90; Vegetius, *Epitoma* 2.22; Josephus, *BJ* 5.2.1.Domaszewski 1885, 6-7; Isaac 1995, 28.

[159] Domaszewski 1908, 44; Alexandrescu 2010, 52. *Contra* Morizot 1995, 241-243, who claims that the two terms should be separated, *tubicen* and *princeps*, and for considering the latter as part of civilian sphere.

[160] Alexandrescu 2010, 52.

[161] Alexandrescu 2010, 67-69.

[162] Domaszewski 1908, 24; Durry 1968, 100. *Contra*, Domaszewski-Dobson 1967, xv: 39 *tubicines* and 36 *cornicines.*

[163] Speidel 1976, 160-161.

[164] Tacitus, *Hist.* 1.38.

[165] Speidel 1992a, 131-132.

[166] AE 1902, 147a.

[167] AE 1902, 147b.

[168] Domaszewski 1908, 44-45.

[169] Speidel 1992b, 136.

[170] *Epitoma* 2.10.

the hierarchy of command in the legion, is charged with coordinating medical personnel, with managing the sick and the wounded and with medical financial issues. In the list of *immunes* from Tarruntenus Paternus (one of the sources of Vegetius) we find medical staff: *optio valetudinarii, medicus, capsarius, veterinarius. Optio valetudinarii* is charged with coordinating the activity of the legionary hospital, this is the same for the cohorts in Rome.[171]

Medici are the most frequently attested in epigraphs. There are several ranks and types of *medici.* The most important *medicus* is a senior one, the leader of the medical staff, most often of Greek origin. He is assimilated to an officer, sometimes to a centurion - *medicus ordinarius.*[172] The other type of *medicus* is obviously inferior hierarchically, sometimes mentioned as *miles medicus.* He is obviously *immunis*, has basic medical skills but enough to treat any medical problem. There are specialized physicians, *medicus chirurgus* and *medicus clinicus*, both in the cohort *VI Praetoria*, or *medicus ocularius* for *classis Britannica* as well as veterinary specialists – *medicus veterinarius* or *medicus pequarius.* The simple *pequarius* is an inferior rank, a sort of nurse. We do not know for certain how many of them were in a legion, it was suggested one for each *centuria*, in any case a large such staff was needed. The most important was this common staff, activating in hospitals, charged with taking care of all the wounded. It is obvious that most of them were highly qualified, of Greek origin and their discoveries and accomplishments are mentioned in ancient sources.[173]

If the highest rank, the *medicus, medicus legionis* or *medicus ordinarius* is *principalis, miles medicus* or *capsarius* are *immunes.*[174] Still the staff of the hospital is the best known. The *optiones valetudinarii*, as *immunes*, are known from the *Digest. Capsarii,* a sort of medical orderlies or nurses, their names deriving from *capsa* – their bag of bandages. *Librarii* and *discentes capsariorum* assist them.[175]

From the category of technicians that are *immunes*, especially in constructions, we find: *architectus*, who seems to be ranked above *custos armorum, discens architectum* and *mensor agrarius*,[176] charged with mapping the *territorium legionis*. Legionary *mensores* are found in several inscriptions, some of them revealing the nature of their function: the *lustrum* of the *territorium legionis*.[177] They can remain as *evocati* or in *officium praefecti Aegypti*. They are written down on the *immunes* list in the *Digest* but their status, just like the status of other technicians, is uncertain.[178] We find attested in several inscriptions however the exact number of *mensores* in a legion. At Lambaesis the number of nine *mensores* of the legion *III Augusta*[179]is a certain fact but at Viminacium the number is 11, probably two for the first cohort. Sometimes they are called *librarii*, probably because the change in profession. Some tasks for the military *mensores* have been proven: drawing and measuring the camp, dividing the *territorium legionis* and the public domain, establishing the centuriation of military colonies, drawing the roads and the *limites.*[180] The unusual place of technicians in the army, especially *architecti*,[181] seems to indicate they were specialized officers with the rank of centurion.[182]

Soldiers tasked with food distribution, with the operation of the *fabrica, horologium*,[183] as well as *equites legionis*[184]can be added to this list.

The career of the legionary

Sources detailing the careers of the legionaries are more numerous and more complex than those detailing the careers of soldiers stationed in Rome, but the longest careers include only four posts, although the number of posts attested as a whole is huge, that this why the structure of the career might seem complicated. Very few can be said about posts ranked below *tesserarius.* Among them only *eques, armorum custos* and the clerks *librarius* and *exactus* are attested. Soldiers serving as clerks were promoted to *tesserarius* or *armorum custos*. Most soldiers probably start up as an *immunis* clerk and then advance in the tactical ranks, to *tesserarius.*[185]

Tactical ranks do not come together as definite as is the case for the cohorts in Rome. The three posts, *tesserarius, optio, signifer* do not always appear in this order. No soldier held the posts *optio centuriae* as well as *optio ad spem ordinis*. In theory the posts of *optio* and those ranked above it were held by soldiers who received double payment. Direct promotion to centurion was possible from any of these posts although, in practice, it comes only after gaining sufficient experience.[186]

There is a number of promotions based on which, one can determine the structure of careers for the *principales* serving in the staff: *frumentarius – beneficiarius consularis, beneficiarius consularis – speculator, speculator – commentariensis, commentariensis – cornicularius consularis, cornicularius consularis – centurio*. When considering the personnel serving under

[171] Davies 1970, 86.
[172] Detailed discussion about this rank in Davies 1969.
[173] Davies 1970, 87-88.
[174] Junkelmann 1986, 110.
[175] CIL VII 1144 – *medico duplic(ario) collegae eius*. Domaszewski 1908, 45. For detailed discussions regarding medical service in the Roman army see Davies 1969 and 1970, Webster 1969, Le Bohec 1994, Speidel 1992, more recently Baker 2004 and the chapter on centurions.
[176] It seems that his duties as professional topographer were limited to placing the campsite of a cohort and probably ensuring its wheat rations as *mensor frumenti*; Arnaud 1995, 251-156.
[177] *VII Claudiae* – CIL III 8112, Viminacium.
[178] In CIL VIII 2564 they appear, alongside other acknowledged *immunes*, on a list of the *duplarii.*
[179] AE 1904, 72.
[180] Sherk 1974, 546-549, 551-558. Centurions or superior officers help them in fulfilling their tasks.
[181] Evans 1994, 147-148. Most are Greek civilians, Pliny, *Ep.*, X.
[182] Stoll 2001, 367; Speidel MA 2001, 53-54.
[183] See recently the forthcoming Cupcea 2014, forthcoming.
[184] Domaszewski 1908, 46-48; Breeze 1969, 53-54.
[185] Breeze 1974, 267.
[186] Breeze 1974, 268-269.

the legion's legate we notice that *beneficiarius legati* was promoted to *commentariensis*, obviously in the governor's staff, although he is below the *beneficiarius consularis*, then to *cornicularius legionis*, finally the post *optio ad spem ordinis* giving all of them access to the rank of centurion.[187]

The posts inside the *centuria* (*optio centuriae*, *optio ad spem ordinis*, *signifer*) were directly connected to the superior ones of *principales*. Promotions from *optio* to *beneficiarius consularis*, from *signifer* to *cornicularius legionis*, from *beneficiarius legati*, *cornicularius legionis* and *commentariensis* to *optio (ad spem ordinis)* are attested. *Signifer* promotes also to *aquilifer*, after that it is usual to promote to centurion.[188]

Apparently there is no specific career a legionary can follow in order to become centurion, the hierarchy being dictated by seniority, because there is no career that has more than four posts and no career with more than two similar posts. The soldiers chosen for advancement were promoted on vacant posts. If we consider only careers with more than two posts, including that of *optio*, only eight posts remain to be considered.[189] Six of these held the post of *optio*, seemingly having only military duties and an administrative post, either *signifer* or a superior post in the staff. The idea or the rule was that a future centurion had to be trained in both aspects of the army – the military and the administrative one. It does not matter when the *optio* post was held but if it was held after the superior post in the staff than the holder is called *optio ad spem ordinis* and is put on a waiting list for promotion to centurion.[190]

It was possible to advance to centurion from a wide variety of posts, from *eques* to *cornicularius consularis*, although with exception of *eques* all other posts were held by soldiers who receive double payment. For the 1st century we have attested the promotion to centurion from the posts of *signifer, aquilifer* and *eques*. Promoting *eques* to legionary centurions is probably connected to the Republican promotion system and towards the end of the 1st century they are promoted as centurions only in auxiliary troops. In the 2nd century the *aquilifer* is still promoted, alongside the experienced *signifer*, *optio* and *optio spei*. During the rule of Marcus Aurelius the first *cornicularius consularis* is promoted to centurion, indicating a change of practice that will eventually become the norm in the 3rd century. Promotions are probable from *cornicularius legionis*, two from *beneficiarius consularis*, but uncertain, and one from *frumentarius*, internal, in the *castra peregrina. Optiones* are still promoted to centurions but without the conditions to be *optio spei*, the last reference regarding this post dating from AD 218.[191] One can notice that in the 3rd century centurions do not have military training comparable to their predecessors from the 2nd century due to the separation of the two parts of the army and promotion from *principalis* to *militia equestris* becomes possible in legions and praetorian cohorts.[192]

Along with the changes in the promotion system of centurions the career of legionary *principales* changes also. In the 1st century the posts of *eques*, *optio*, *signifier* and *aquilifer* are emphasized and *cornicularius* and *beneficiarius* appear very rarely in pre-Flavian careers. All these posts maintain their importance in the 2nd century and apparently, with the exception of *aquilifer*, in the 3rd century also. The posts in the staffs of superior officer receive greater and greater importance and dominate careers. The first promotion from a post in the staff to centurion in made by *optio ad spem ordinis*, under Trajan. From Marcus Aurelius onwards centurions promote directly from *cornicularius consularis*, thus opening the way for promotions exclusive from posts in the staff, in the 3rd posts.[193]

As for the problem of duration of service before promotion to centurion the careers we know seem to indicate a variation from 14 to 24 years. Promotion could occur at anytime, but most seem to take place after 13-20 years of service; as far as praetorians are concerned it occurred after 15-16 years. After the 20th year of service the chances for promoting to centurion were reduced. Soldiers made petitions to be promoted to centurion at any moment in their career as *duplicarius*. Centurions had held 4-5 previous posts, each one for about three years, those that did not enjoy the perspective of promotion held them for longer periods and the last post was held for a very long time, until discharge.[194]

Some soldiers held a post of *immunis* in the beginning, maybe as a minor functionary, then were promotion in a post in the *centuria*, especially *tesserarius* but also *optio*, after approximately 6 years of service. From the posts of *optio* and *signifer* one could promote directly to centurion or, more frequently, to a superior staff post, culminating with *optio ad spem ordinis*. Promotion to centurion probably takes place between the 14th and the 16th years of service.

---

187 Breeze 1974, 269.

188 Breeze 1974, 269.

189 Breeze 1974, 270: L. Val. Proculus – CIL III 12411=ILS 2666b; M. Sabidius Maximus – AE 1037, 101; Petronius Fortunatus – CIL VIII 217=11301=ILS 2658; Peregrinus – CIL V 7004; Amandianus – CIL III 11135=ILS 4311; Mommius Cattianus – AE 1951, 194; Fl. Maximus – ILS 8880; Iul. Ingenus – CIL VIII 12128.

190 Breeze 1974, 269-270.

191 CIL III 3445.

192 Breeze 1974, 272. See also Le Bohec's study on *petitores*, Le Bohec 2012.

193 Breeze 1974, 272-273.

194 Breeze 1974, 273-276.

I.2.3. The lower hierarchy of legionaries from Dacia

XIII Gemina

Of all the military units in Dacia the legion *XIII Gemina* has the highest number of attestations of petty officers. An impressive number of 143 *immunes* and *principales* are mentioned in the whole province as belonging to the legion. Of these the vast majority obviously belong to the staff of the consular governor (75). The other *officia* are also represented: of the legate of the legion (14), of the tribunes (3) and even of the prefect of the legion (4). The list also contains 15 tactical ranks and only 11 *immunes* that activate in the *centuriae*. As a whole, two thirds (107) hold *principales* ranks/posts and the rest (36) are just *immunes*.

Officium consularis. The most important bureaucratic apparatus in the whole province was ample and had a complex structure, offering the governor the necessary abilities to coordinate all provincial activities. When analyzing the legion at Apulum we see that it is represented by a number of 75 persons, 62 of them being *principales*. The most numerous members of this staff are the *beneficiarii* (32). There are also attested 13 individuals and a collective of *speculatores*, six *quaestionarii*, three *frumentarii* and three individuals plus a collective of *stratores*. From the *immunes* serving in this *officium* there are attested one *adiutor offici corniculariorum*, five *librarii* and seven *exceptores*.
The most important position of *principalis* to be found in the career of a legionary that served in an *officium* is that of *cornicularius consularis*. It is the post at the head of all bureaucrats that form the *officium consularis*. Promotion from this post is directly to centurion in a legion and is only a formality. We have three such *cornicularii* belonging to the legion *XIII Gemina* as well as a collective, all attested at Apulum.

Nos. **1** and **2** are present on a petty officers list, members of *officium consularis*, from Apulum where they are inscribed on three columns as taking part in the building of *schola speculatorum*, for the health of the emperors Severus and Caracalla, of the *Caesar* Geta and under the orders of the governor Mevius Surus (AD 198-199). Immediately after the three legionary centurions mentioned at the top of the list are the two *cornicularii*. Further down the monument is badly damaged permitting only the partial recognition of those *principales* who served in leading posts.[195]

On a marble fragment from a votive or honorific monument three collectives of the highest functionaries of the consular staff are mentioned. The first are *cornicularii*. They are probably the ones who set up the monument, maybe as a gesture of devotion for the emperor (Caracalla?).

The same epigraph features in the second line the *commentarienses*, coauthors of this gesture of devotion. It is the only attestation in Dacia of one of the most important posts from the *officium consularis*, tasked with legal duties next to the governor.

Next comes the rank of *speculator*, the personal courier of the governor and a man of confidence even in relation to the imperial authority. We find attested 13 individuals and a collective of *speculatores*, as well as the headquarters of their organization from inside the *praetorium* of the governor: *schola speculatorum*.

Number **5**, *speculator legionis XIII Geminae Gordianae*, sets up a marble statue base for Apollo after following his command and rebuilding the fountain of the eternal god. On the above mentioned list of petty officers from Apulum there are attested a number of 10 *speculatores*, nos. **6-15**, all of them from the legion *XIII Gemina*, and serving under governor Mevius Surus in AD 198-199. According to I. Piso if on the last two columns there are mentioned ten *speculatores* from each legion of the province and assuming that the first one contains those from *XIII Gemina,* it means that out of the ten *speculatores consularis* eight are unknown.[196]

Finally, on the same fragment of honorary inscription (CIL III 7794b=IDR III/5 435), *speculatores* are mentioned as participating in the dedication, following after the *cornicularii* and the *commentarienses.*

An impressive number of *beneficiarii* are attested from the legion *XIII Gemina*, 32. More than half of them are attested Apulum, they were not transferred in the province for some special mission. Other locations where they are also present are: Sarmizegetusa, Micia, Alburnus Maior, Ampelum or Samum. Out of all these locations we can state that those detached in *stationes* are the ones from Micia, Alburnus Maior and Samum.

A very important character among the ones from Apulum is no. **19**. His first mention is on an altar/statue base, dated under Severus, from Apulum, dedicated to Diana *Augusta*, as veteran of the *XIII Gemina* legion, former *beneficiarius* of the consular, presently decurion of the Sarmizegetusa *colonia*. Later on, during the reign of Severus Alexander, we find him on an honorific monument set up by his sons at Sarmizegetusa, all of them being exceptional soldiers, all of them officers, one even praetorian and it is from this monument that we find out he pursued a complete municipal career, until the *IIvirate*. We are dealing with a very appreciated and exceptionally honored family from Sarmizegetusa, led by a professional soldier and a significant member of the local elite.

Other 14 *beneficiarii* are attested as detached in different places in Dacia, on duty or at home. Four are at Sarmizegetusa, two veterans and city officials and two active soldiers; one of them being also attested at Apulum

[195] For such an attempt see Cupcea 2008, *passim*.

[196] See also Cupcea 2008, 270-277.

No. **33** is one of the sons of no. **19**, also a *beneficiarius*, honoring his father by setting up the discussed monument. No. **35** appears on a distinct funerary monument. The first part of the epigraph mentions that him, veteran, former *beneficiarius* of the consular, decurion of the colony, burying his wife. The second part is his own epitaph that reveals he also held the municipal office of *quaestor* and *IIvir*, and that he has two sons Valerius Rufinus and Valerius Proculeianus who set up the monument.

At Micia we find three consular *beneficiarii*, nos. **36-38**, probably revealing that a *statio* was established here. Also it is possible there was a *statio* at Alburnus Maior, because here we have one of the largest gold mining operations in the whole Empire and we find attested three *beneficiarii*, nos. **40-42**. Nearby, at Ampelum, we find a *beneficiarius* of the consular, no. **38** who dedicates a votive altar to Iupiter *optimus maximus* Dolichenus, for the health of the perpetual emperor, Gordian III. His presence here cannot be certainly connected to the office of the *aurariarum* procurator, as suggested by the authors of the IDR, because there are different authorities that the two relate to. Certainly he represents the authority of the consular in this very important place and we can only guess as to the nature of his activity here.[197]

Another *statio*, certain and documented, is the one from Căşei-Samum. The legion *XIII Gemina* is represented here by at least three consular beneficiaries, nos. **44-46**. The last one sets up a votive altar for the goddess Nemesis *Regina*, both as a legionary and as an *aedile* of the *colonia* Napoca; on this altar he describes his mission and it is dated based on the names of the consuls in AD 239.

No. **43** sets up a votive altar for Iupiter *optimus maximus*. The altar is reused in the construction of the late feudal fortification from Cluj-Mănăştur, thus we cannot precisely determine its place of origin. Still the existence of a *statio* at Napoca cannot be assumed based only on one evidence,[198] especially when considering the fact that these beneficiaries were often dispatched through the cities in the province with customs and tax-collecting duties.

This is the last item in our last list of *beneficiarii*. The next hierarchical post is *questionarius*. Again we are dealing with a post reserved for the staff of governors, with police, interrogation and investigation duties. We have a number of six discovered *quaestionarii*, all at Apulum, on the same list, an honorific inscription probably marking their *collegium*.

Frumentarii are also part of the *officium consularis* and they have similar or related duties to those of the *speculatores*, although they do not probably serve on guard duty. There is only one *frumentarius* in Dacia, at Sarmizegetusa, from the legion *XIII Gemina*, No. **55**, who is one of the sons of the beneficiary from No. 19, previously discussed; at the same time he is decurion of the colony, a highly-honored position.[199]

The *frumentarii* of the Roman army from Dacia are attested in other provinces of the Empire, while travelling for the different missions they received. One of them, No. **55'** *frumentarius legionis XIII Geminae*, dies while serving in Rome and is buried there by his colleagues.[200] No. **55''** also belongs to legion *XIII Gemina Severiana*, and is sent in a *statio* on Via Appia along with a colleague from *VIII Augusta*.[201]

Strator, a member of the legionary guard of the governor, is also part of his staff. There are 5 individuals and a collective of *stratores* all discovered at Apulum.

No. **56** has one of the few military careers discovered in Dacia. On one altar dedicated to Iupiter *optimus maximus,* our character, who originates from the Velleia tribe, in Aquileia, states he is centurion of the legion *XIII Gemina*, promoted from the post of *strator* of the governor, and at the same time of the legate of the legion L. Iulius Proculus (approx. AD 153-156). The collective of *stratores* from the legion *XIII Gemina* sets up a votive altar to Iupiter *optimus maximus,* for the health of the governor C. Hasta (approx. AD 190), through the legate of the legion, Proculus. Nos. **59-60** are found on the most recently discovered list of soldiers from Apulum, however they could also be in the service of the legionary legate.[202]

The last 13 members of the *officium consularis* are *immunes*, all of them secretaries. We are referring to one *adiutor offici corniculariorum*, five *librarii* and seven *exceptores*, from different cities of Dacia.

No. **61**, *adiutor offici corniculariorum* from the legion in Apulum is the son of the centurion from the *numerus Palmyrenorum O?*, P. Aelius Septimius Audeo from Sarmizegetusa, for whom he along with the rest of the family sets up the funerary monument. *Librarii* usually set up altars in honor of Minerva. So too does a group of six *exceptores consularis*, nos. **68-69**, who dedicate an altar to Minerva *Augusta*. The names of only three of them cannot be reconstructed.[203] Finally, another *exceptor* from Napoca, no. **67**, deceased at the age of 20, is mentioned in connection to an *actarius* from the cohort *I Alpinorum* (?), proving the relations between these soldiers-secretaries, that transcend the limits of the military unit, most likely in the *officia.*

The other staff made up of legionaries from the *XIII Gemina* is the ***officium legati legionis***. It is far less documented than the *officium consularis*, presenting 13 characters, grouped on five posts. Out of these the only

[197] Hirt 2010, 159-161, 165, 259.
[198] *Contra* Ott 1995, 289.

[199] See also Cupcea 2009, 309-310.
[200] CIL VI 3356=IDRE I 25.
[201] CIL VI 230=36748. Details about the *frumentarii* and Dacia in Cupcea 2009.
[202] Moga, Drâmbărean 2007, 176-177. See also Cupcea 2013a.
[203] Berciu, Popa 1964, *passim*.

posts of *principales* are those of *beneficiarius legati* and possibly *actarius*, the rest being of *immunes* secretaries.

We know a number of seven beneficiaries; only one of them, a veteran is encountered far from the headquarters of the legion, at Apoldu de Sus: it is no. **80** who probably had properties here.

The next six *immunes* all are different types of secretaries, some of them coordinating the activities in certain bureaus, discovered at Apulum and Ampelum.

No. **81**, *actarius legati legionis XIII Geminae*, sets up along with his wife an altar dedicated to the goddess Iuno *Semlia.* It is from this post that he coordinates the *acta diurna* of the legion, regarding the daily schedule. Only one of the six *librarii legionis* that are to be discussed next is not present in the legionary fortress from Apulum. An unknown *librarius*, no. **86**, sets up along with his colleagues *commanunculis*, a votive marble plaque dedicated most likely to the centurions of the legions, dated in AD 227, 232 or 253. It might highlight the secretary's desire to reach the post of centurion.

Of the ***officium tribuni legionis*** there are only three members attested, all of them *immunes*, on two lists of soldiers. Their names are unknown and they occupy the posts of *beneficiarius* and *secutor*.

The last of the *officia* of the legion *XIII Gemina* is ***officium praefecti legionis***. From this office we know four members on four posts; in the case of one of them we also have a short career.

No. **91** dedicates a votive altar for Mars *conservator*, for the health of the emperors Severus and Caracalla (AD 198-209). He has a simple career, in the *officium rationum*, the staff of the prefect of the legion, promoting from *adiutor offici rationum* to *cornicularius*.

Out of the 38 ***principales*** the tactical ranks are very well represented along with the higher ones, of *candidatus* and those of *duplicarius* and *eques*. Besides them the ranks of *imaginifer* and that of *conductor armamentaria* are also mentioned. Only three of them did not leave any evidence behind at Apulum.

The rank of *candidatus*, waiting for a promotion to centurion, is found six times on the same list of soldiers mentioned above, nos. **95-100**. The only *imaginifer*, no. **101**, dedicates a votive mithraic relief depicting the slaying of the bull.

The **tactical ranks** are attested 15 times: four times *optio centuriae,* four times *signifer,* four times *tesserarius*, two times *optio signiferorum* and once *optio fabricae*.

The close connection between the *optio* and his *centuria* is illustrated by the dedication for the *genius centuriae* made by no. **102.**[204] No. **103** is grateful to the nymphs of Germisara who have healed him. A special category of *optio* is mentioned, *optio fabricae (cos?)*, no. **108**, probably a post lower than *optio centuriae.*[205] The other very important tactical rank, *signifer*, is mentioned four times in the legion *XIII Gemina.* Also, two *optiones signiferorum*, meaning the leaders of the flag-bearers' association, are present. From the *tesserarii* no. **115** comes from a military family, one of his brothers being *duplicarius.*

A number of ten *duplicarii* are mentioned in the legion *XIII Gemina*. We are inclined to believe that they are *principales* of the *equites legionis*, although they are not widely mentioned. Another opinion is that they would not refer to a different pay but to an increase in the food ration.[206] *Duplicarii* from nos. **123-128** are present on the list of soldiers published in 2007.

We must also include in the cavalry registry the four *equites legionis* mentioned in the legion *XIII Gemina,* because their status as *principales* is obvious. Nos. **131** and **132** come from outside Dacia, from Dalmatia, respectively Siscia, indicating a possible recruitment of legionary riders from areas that have such a background.[207]

Finally, the last rank in between *immunes* and *principales*, is *conductor armamentarii*, nos. **117-118.** Found twice on the same votive relief, the post can be a variant of *custos armorum*, but it seems to be much more than just that, a petty officer responsible for the whole arsenal of the legion.

To conclude only 11 ***immunes*** that do not activate in any *officium* can be accounted for from the legion *XIII Gemina.*

Out of them three are *custodes armorum*, three are military musicians, - from the category most often encountered – *tubicines*, no. **139** is one of the few *agrimensores* found in the Empire, despite the fact that he is a very important technician and finally three are simply *immunes*. No. **141** originates from the vexillation that built the fort of Micia.

Finally, the interesting case of no. **143** could be the single accurate epigraphic attestation of a *horologiarius* in the Empire.[208]

[204] Other dedication of this type at Potaissa, see Speidel, Dimitrova 1978 and Bărbulescu 2004.
[205] On the *fabri*, see recently TVind IV-1 and IV-2.
[206] Sander 1959, 247.
[207] Although if we apply the theory stating that a soldier was enlisted as *pedes*, then promoted to *eques*, origin loses its complete importance on this aspect. Still it is probable that the situation of the *equites legionis* was somewhat different.
[208] For the discussion, see Cupcea 2014, forthcoming.

	Name	Rank/Post	Source	Place
		Officium consularis		
1	Aelius Valerius	*cornicularius*	CIL III 14479=IDR III/5 426	Apulum
2	Antonius Va[---?]	*cornicularius*	CIL III 14479=IDR III/5 426	Apulum
3	Celerinus Valerianus	*cornicularius*	CIL III 1106=IDR III/5 264	Apulum
	Collective	*cornicularii*	CIL III 7794b=IDR III/5 435	Apulum
	Collective	*commentarienses*	CIL III 7794b=IDR III/5 435	Apulum
4	P. Aelius Valerianus	*speculator*	IDR III/5 721	Apulum
5	Ulpius Proculinus	*speculator*	CIL III 990=IDR III/5 31	Apulum
	Collective	*speculator*	CIL III 7794b=IDR III/5 435	Apulum
6	Caius	*speculator*	CIL III 14479=IDR III/5 426	Apulum
7	Cocceius	*speculator*	CIL III 14479=IDR III/5 426	Apulum
8 to 15	8 x *Unknown*	*speculator*	CIL III 14479=IDR III/5 426	Apulum
16	Statius Alexander	*speculator*	AE 2007, 1198	Apulum
17	*Unknown*	*bf cos*	CIL III 1189=IDR III/5 453	Apulum
18	Claudius Alexander	*bf cos*	CIL III 1190=IDR III/5 454	Apulum
19	C. Iulius Valerius	*bf cos*	CIL III 7742=IDR III/5 60; IDR III/2 113	Apulum, Sarmizegetusa
20	Caecillius Saturninus	*bf cos*	CIL III 7755=IDR III/5 138	Apulum
21	Iulius Rufus	*bf cos*	CIL III 1039=IDR III/5 138	Apulum
22	Iulius	*bf cos*	IDR III/5 244	Apulum
23	*Unknown*	*bf cos*	IDR III/5 291	Apulum
24	L. Aelius Silvanus	*bf cos*	CIL III 1185=IDR III/5 378	Apulum
25	Longinus	*bf cos*	IDR III/5 594	Apulum
26	L. Arrius Probus	*bf cos*	CIL III 1080=IDR III/5 201	Apulum
27	Vindex	*bf cos*	CIL III 1059=IDR III/5 178	Apulum
28	Aurelius Ianuarius	*bf cos*	IDR III/5 135	Apulum
29	Sextus Clemens	*bf cos*	CIL III 1040=IDR III/5 142	Apulum
30	Gallicanus	*bf cos*	CIL III 7736=IDR III/5 27	Apulum
31	P. Aelius Fronto	*bf cos*	CIL III 1091=IDR III/5 233	Apulum
32	Terentius Marcianus	*bf cos*	CIL III 14474=IDR III/5 295	Apulum
33	C. Iulius Valerianus	*bf cos*	AE 1933 248=IDR III/2 113	Sarmizegetusa
34	Ulpius Maximianus	*bf cos*	IDR III/2 239	Sarmizegetusa
35	L. Valerius Rufus	*bf cos*	CIL III 1485=IDR III/2 452	Sarmizegetusa
36	Claudius Saecularis	*bf cos*	AE 1933, 9=IDR III/3 86	Micia
37	Priscinus	*bf cos*	AE 1933, 9=IDR III/3 86	Micia
38	T. Flavius	*bf cos*	CIL III 7859=IDR III/3 124	Micia
39	Aurelius Caius	*bf cos*	AE 1971, 381=IDR III/3 297	Ampelum
40	Q. Marius Proculus	*bf cos*	AE 1990, 827=ILD 359	Alburnus Maior
41	C. Calpurnius Priscianus	*bf cos*	AE 1990, 828, 840 =ILD 360, 373	Alburnus Maior
42	C. Iucundius Verus	*bf cos*	AE 1990 837=ILD 370	Alburnus Maior
43	P. Aelius Fabianus	*bf cos*	ILD 544	Napoca
44	Aurelius Rufinus	*bf cos*	ILD 774	Samum
45	Val. Vibius Valerianus	*bf cos*	CIL III 823	Samum
46	Valerius Valentinus	*bf cos*	CIL III 827	Samum
47	Helvius	*bf cos*	AE 2007, 1198	Apulum

48	Aurelius ?	*bf cos*	AE 2007, 1198	Apulum
49	Longius Orestus	*quaestionarius*	CIL III 7803=IDR III/5 459	Apulum
50	Iulius Victorinus	*quaestionarius*	CIL III 7803=IDR III/5 459	Apulum
51	Ulpius Valentinus	*quaestionarius*	CIL III 7803=IDR III/5 459	Apulum
52	Iulius Seneca	*quaestionarius*	CIL III 7803=IDR III/5 459	Apulum
53	Aelius Caesianus	*quaestionarius*	CIL III 7803=IDR III/5 459	Apulum
54	Aelius Valerius	*quaestionarius*	CIL III 7803=IDR III/5 459	Apulum
55	C. Iulius Carus	*frumentarius*	AE 1933 248=IDR III/2 113	Sarmizegetusa
55'	M. Aurelius Berulus	*frumentarius*	CIL VI 3356=IDRE I 25	Rome
55''	M. Aurelius Sophaenetus	*frumentarius*	CIL VI 230=36748	Italy
56	C. Silius Crispinus	*strator* *centurio*	AE 1977 653=IDR III/5 166	Apulum
57	Aurelius Mucianus	*strator*	CIL III 7786=IDR III/5 383	Apulum
58	M. Ulpius Respectus	*strator*	AE 1930, 5=IDR III/5 61, 180	Apulum
	Collective	*stratores*	AE 1983, 801=IDR III/5 137	Apulum
59	Tertius Iustus	*strator*	AE 2007, 1198	Apulum
60	Iulius	*strator*	AE 2007, 1198	Apulum
61	Ael. Sept. Romanus	*adiutor off cornic*	CIL III 1471=IDR III/2 366	Sarmizegetusa
62	P. Aelius Propincus	*librarius*	AE 1982, 826a=IDR III/5 482	Apulum
63	Aelius Sabinus	*librarius*	CIL III 14215[16]=IDR III/5 266	Apulum
64	M. Iulius Victorinus	*librarius*	AE 1982, 825=IDR III/5 544	Apulum
65	P. Helvius Primanus	*librarius*	CIL III 1318=IDR III/3 354	Ampelum
66	Quintillianus	*librarius*	CIL III 6246	Gherla
67	*Unknown*	*exceptor*	AE 1987, 841=ILD 561	Napoca
68	Florus	*exceptor*	AE 1964, 193=IDR III/5 263	Apulum
69	Severus	*exceptor*	AE 1964, 193=IDR III/5 263	Apulum
70	Valens	*exceptor*	AE 1964, 193=IDR III/5 263	Apulum
71 to 73	3 x *Unknown*	*exceptor*	AE 1964, 193=IDR III/5 263	Apulum
		Officium legati legionis		
74	Geminius Aelianus	*beneficiarius*	CIL III 1026=IDR III/5 96	Apulum
75	Crispius Saturninus	*beneficiarius*	CIL III 1031=IDR III/5 111	Apulum
76	Valerius Senecio	*beneficiarius*	AE 1980, 746=IDR III/5 590	Apulum
77	Sempronius Valentinus	*beneficiarius*	CIL III 1050=IDR III/5 163	Apulum
78	Aurelius	*beneficiarius*	CIL III 14480=IDR III/5 449	Apulum
79	M. Memmius Longus	*beneficiarius*	CIL III 7740a=IDR III/5 15	Apulum
80	*Unknown*	*beneficiarius*	CIL III 969=IDR III/4 13	Apold de Sus
81	Iulius Alexander	*actarius*	CIL III 7753 = IDR III/5 108	Apulum
82	M. Aurelius Antoninus	*librarius*	CIL III 1317 = IDR III/3 344	Ampelum
83	*Unknown*	*librarius*	CIL III 1189 = IDR III/5 453	Apulum
84	M. Marcellus	*librarius*	CIL III 1166 = IDR III/5 391	Apulum
85	M. Mociuncius Valentinus	*librarius*	CIL III 1194 = IDR III/5 556	Apulum
86	*Unknown*	*librarius*	IDR III/5 406	Apulum
87	Ti. Iulius Novianus	*librarius*	CIL III 1105 = IDR III/5 261	Apulum
88	*Unknown*	*beneficiarius*	CIL III 1190=IDR III/5 454	Apulum
89	*Unknown*	*beneficiarius*	AE 2007, 1198	Apulum
90	*Unknown*	*secutor*	CIL III 1190=IDR III/5 454	Apulum
		Officium praefecti legionis		

91	P. Aelius Rufinus	*cornicularius* *adiutor offici rationum*	AE 1955, 1290 = IDR III/5 250	Apulum
92	P. Aelius Proculinus	*beneficiarius*	AE 1995, 1291 = IDR III/5 699	Apulum
93	L. Veratius Speratus	*beneficiarius*	CIL III 1056 = IDR III/5 173	Apulum
94	C. Iulius Sabinus	*libr a rationib*	CIL III 1477 = IDR III/2 419	Sarmizegetusa
		Principales		
95	Septimius Celsus	*candidatus*	CIL III 1190=IDR III/5 454	Apulum
96	Septimius Alexander	*candidatus*	CIL III 1190=IDR III/5 454	Apulum
97	Septimius Nepos	*candidatus*	CIL III 1190=IDR III/5 454	Apulum
98 to 100	3 x *Unknown*	*candidatus*	CIL III 1190=IDR III/5 454	Apulum
101	M. Ulpius Linus	*imaginifer*	IDR III/5 290	Apulum
102	Ti. Cl. Mansuetus	*optio*	CIL III 1015=IDR III/5 79	Apulum
103	M. Aur. Mosianus	*optio*	AE 1993, 1341=ILD 319	Germisara
104	*Unknown*	*optio*	CIL III 14478=IDR III/5 395	Apulum
105	*Unknown*	*optio*	CIL III 7792=IDR III/5 407	Apulum
106	Aurelius Marius	*optio signiferorum*	CIL III 1124=IDR III/5 247, 293	Apulum
107	M. Ulp. Flavinus	*optio signiferorum*	CIL III 1202=IDR III/5 605	Apulum
108	Ael. Proc(ulinus?)	*optio fabricae cos?*	AE 2007, 1198	Apulum
109	C. Iulius Marcianus	*signifer*	IDR III/5 282	Apulum
110	*Unknown*	*signifer*	AE 1983, 814=IDR III/5 669	Apulum
111	Vibius Rufinus	*signifer*	CIL III 1192=IDR III/5 521	Apulum
112	C. Val. Ingenuus	*signifer*	IDR III/3 15	Călan
113	Por(cius?) Iulianus	*tesserarius*	CIL III 1189=IDR III/5 453	Apulum
114	Q. Aurelius ?	*tesserarius*	AE 1983 813=IDR III/5 503	Apulum
115	L. Ael. Candidinus	*tesserarius*	CIL III 8034=IDR II 352	Romula
116	*Unknown*	*tesserarius*	Moga, Drâmbărean 2007	Apulum
117	Turranius Marcellinus	*conduct. arm.*	CIL III 1121 = IDR III/5 285	Apulum
118	Ant. Senecio Iunior	*conduct. arm.*	CIL III 1121 = IDR III/5 285	Apulum
119	Aurelius Senecio	*duplicarius*	CIL III 1189 = IDR III/5 453	Apulum
120	Pomponius Victorinus	*duplicarius*	CIL III 1067 = IDR III/5 188	Apulum
121	*Unknown*	*duplicarius*	CIL III 1190 = IDR III/5 454	Apulum
122	*Unknown*	*duplicarius*	CIL III 1190 = IDR III/5 454	Apulum
123	Valerianus	*duplicarius*	AE 2007, 1198	Apulum
124	*Unknown*	*duplicarius*	AE 2007, 1198	Apulum
125	Genialis	*duplicarius*	AE 2007, 1198	Apulum
126	Valentinus	*duplicarius*	AE 2007, 1198	Apulum
127	Bassinus	*duplicarius*	AE 2007, 1198	Apulum
128	Valerianus	*duplicarius*	AE 2007, 1198	Apulum
129	P. Aelius Secundinus	*eques*	AE 1980, 738 = IDR III/5 377	Apulum
130	Antonius Valens	*eques*	AE 1983, 800 = IDR III/5 450	Apulum
131	Sextus Victor	*eques*	CIL III 1200 = IDR III/5 601	Apulum
132	*Unknown*	*eques*	AE 1982, 826b = IDR III/5 624	Apulum
		Immunes		
133	Filomelus	*custos arm.*	AE 1980, 740= DR III/5 529	Apulum
134	D. Iulius Rusonius	*custos arm.*	CIL III 1138=IDR III/5 307	Apulum
135	Ulpius	*custos arm.*	Moga, Drâmbărean 2007	Apulum
136	Varius Firminus	*tubicen*	IDR III/1 274	Sânnicolau Mare

137	C. Sentius Bassus	*tubicen*	AE 1978, 664=IDR III/5 224	Apulum
138	Aurelius Licin(ius?)	*tubicen*	Moga, Drâmbărean 2007	Apulum
139	Aurelius Lucillius	*agrimensor*	CIL III 1189=IDR III/5 453	Apulum
140	M. Aur. Crescentinus	*immunis*	CIL III 1038=IDR III/5 134	Apulum
141	T. Aurelius Arimo	*immunis*	CIL III 12565=IDR III/3 39	Micia (Deva)
142	Pere(grinus?)	*immunis*	AE 2007, 1198	Micia (Deva)
143	M. Ulp. Mucianus	*horologiarius*	IDR III/5 193	Apulum

Table I.1. *Immunes* and *principales* of the legion *XIII Gemina* in Dacia.

V Macedonica. It is a well-known fact that from the military point of view Dacia was born as an appendicle to Moesia Superior and Pannonia, Muntenia and southern Moldavia being connected to Moesia Inferior. Thus the three legions of the province, *I Italica*, *XI Claudia* and *V Macedonica*, were tasked with organizing this territory. During the Trajanic period the legion *V Macedonica*, previously camped at Oescus, appears on several tile stamps in Oltenia and Muntenia. The materials discovered at Drobeta can be related to the construction of the bridge.[209]

After the second Dacian war it remained at Troesmis where it was useful in the confrontations with the *Roxolani* and in the Eastern campaigns from the mid-second century, sometimes being moved completely.[210] It leaves in the Orient with Lucius Verus, returning in Europe for the Marcomannic wars, it is placed from AD 168 in Dacia, under the authority of M. Claudius Fronto. It will remain at Potaissa until AD 271, when it will return to Oescus.[211]

When it comes to the evidence regarding *immunes* and *principales* the legion *V Macedonica* is very well placed with 139 such characters known throughout the whole province. Out of these 27 are members of the *officium consularis* from Apulum, 12 are from the *officium legati legionis*, two are from *officium tribuni*, 60 stand on tactical ranks and 14 are *immunes*.

The registry will start as usual with the soldiers activating in the staff of the consular governor from Apulum. ***Officium consularis*** includes 27 men from the legion in Potaissa, on only five posts, all of them *principales*. The highest-ranking post of this staff found in the *V Macedonica* is that of *speculator*; we found evidence for 11 such persons.

No. **1**, a former *speculator,* is mentioned on a list of veterans, who honor a *sacerdotalis Daciae*, dating back to the 3rd century. The other "half" of the speculators in service of the consular that are mentioned on the building plaque of the *schola speculatorum* from the governor's *praetorium* includes the ten such petty-officers that come from *V Macedonica*; the name of only one of them being decipherable.

Most of the governor's petty officers are *beneficiarii*, a total of 13, most of them (eight) being found in the *statio* from Căşei-Samum, only one being sent to the governor's *praetorium*, in Apulum. A *schola beneficiariorum* is also mentioned at Potaissa.

No. **12**, a veteran, a former beneficiary of the consular, along with a comrade, Serennius Martinianus, a centurion in the legion, dedicates a votive altar, probably to Iupiter *optimus maximus*. This association of the ranks can illustrate the ambitions and perspectives of a petty officer. Similar to the *speculatores*, *beneficiarii* are grouped together in associations, the headquarters being called *scholae*. A collective of beneficiaries of the consular L. Octavius Iulianus, who governed Dacia sometime between AD 200 and 210,[212] dedicate an altar to the *genius* of one of their *schola* in the fortress from Potaissa.[213]

Nos. **22** and **23** present the possibility that there was a *statio* at Drobeta, closely connected with the bridge that was here, a very important transit point and possibly customs.

No. **24** is a very important character from Apulum. A veteran, former beneficiary of the consular, he becomes decurion of the colony from Apulum and makes numerous dedications. First of all he sets up a votive altar in honor of Aesculapius and Hygia and the other gods who lived in that sanctuary, after he recovered his sight, for the health of himself and his family. Secondly he builds a bridge inside the same sanctuary for the health of his kin, following a divine commandment and dedicates the building plaque to Apollo, Diana, Latona as well as to the other gods housed in there. His activity seems to take place sometime during the reign of Commodus.

The next eight beneficiaries of the consular, nos. **14-21** are dispatched in the *statio* from Samum, a place of special importance but whose significance is still difficult to determine exactly. They end the list of beneficiaries of the consular that originate from the *V Macedonica.* They are placed in key positions inside Dacia, having clear missions. First of all most of them populate the *statio* from Samum, as is only normal because of the proximity to the legionary fortress of Potaissa. Then some of the them are in the *statio* from Drobeta where it is obvious that there was a strategic point that needed military surveillance; the soldiers from *V Macedonica* were traditionally sent to serve here following the constructive work of the legion from the beginning of the province.

Only other three posts from the *officium consularis* are mentioned.

No. **27** is *optio praetorii*, from the legion *V Macedonica*, meaning he is the second-in-command of the centurion that effectively leads the *officium consularis*, *princeps praetorii*, or, according to the latest research his whole *praetorium*.[214] One *strator* and one *adiutor offici corniculariorum* are also mentioned.

Officium legati legionis at Potaissa is represented by 11 persons, serving on four posts. As in the case of the governor's staff the *cornicularius* is first in the hierarchy.[215] He is closely followed by a number of five *beneficiarii legati legionis*.

[209] Piso 2000, 213.
[210] Piso 2000, 214.
[211] Piso 2000, 214-218.

[212] Piso 1993, 160.
[213] Bărbulescu 1987, 73, says that they are *beneficiarii* of the legate of the legion. If this were so, normally they would have expressed their gratitude towards him instead of the consular.
[214] Rankov 1999, 19.
[215] Ommited by K. Stauner in his general tables, Stauner 2004, 162-165.

The case of no. **30** is also interesting. He is a beneficiary of the legate Tiro, sets up the funerary monument of his brother, centurion Ferox and had served for 25 years, 18 as *eques legionis*. The legate is identified as T. Calestrius Tiro, a commander of the legion *V Macedonica*, while it is stationed at Oescus.[216]

No. **32** is of African origin, from Castellum Zigal, Numidia. The rest of the four petty officers all are *librarii*.

These are the last to be mentioned as member in the staff of the legion's legate. Along it there was an ***officium tribuni legionis***, attested in the legion *V Macedonica*, in Dacia, by two *beneficiarii*; out of the two no. **40** seems to have been sent away in the *statio* from Drobeta along with other beneficiaries of the consular.

We know very well the ***principales*** who activated in the *centuria* and just as well the *principales* of the legionary cavalry, a total of 28 persons. Besides them two other types of superior petty officers are mentioned.

No. **41** is known as a veteran, *candidatus* of the legion *V Macedonica,* probably on his funerary stele. He waited for a while to be promoted to centurion and probably because there were no vacant posts, retired. Another *candidatus* and a single *imaginifer* are mentioned also.

The **tactical ranks** are very well represented, especially after the reconstruction of a list of *signiferi*, recently discovered at Potaissa.[217]

The reconstruction proposed is somewhat problematic, but basically we are dealing with a marble construction plate of an *exedra* in the *schola signiferorum*, appartently located in the *principia* of the fortress of Potaissa. There are however a few alterations that I would like to make regarding to the published variant. First, it is probable that the inscription would start with an imperial title, maybe that of Gordian III[218] then the fragment mentioning *signiferum* causes a misunderstanding. The author argues that we are dealing with a *discens signiferum*, therefore with an apprentice in standard bearing.[219] However it is highly irregular for such a character to lead the list of officers and in this case their *collegium*. We should see in this an abbreviation of the form *signiferorum*, which clearly belongs to an *optio signiferorum*, the actual leader of the *collegium*, as confirmed also in Apulum,[220] therefore no. **49**. The other fragments of the inscription would show the names of the members of this *collegium*, all of them the *signiferi* in post at this time in *V Macedonica*. If we consider the graphic reconstruction accurate, then we have place for 48 such petty officers, grouped on three columns, of which only 16 are known by name (another six being impossible to reconstruct). Finally, the lower part contains the curator of this work, one Maximus, who could be either a member of this *collegium*, therefore a standard-bearer, or an aspiring member, in which case he is the possible *discens signiferum*.

In any case, we have another attestation of a military association in Potaissa, of again a group of petty officers, this time standard bearers that receive an official headquarters in the most official building of the fortress. They choose to mark this event by a dedication through which all the members that have contributed to the building are known. Even if no actual rank is attached to the remaining names, we can consider them *signiferi*, even if the presence of the *aquilifer* or the *imaginifer* cannot be excluded, as these are the most senior of the legionary standard-bearers and they are only two, therefore a *collegium* of their own would be unlikely.[221]

What does this number can say about the effectives of the legion or the number of *signiferi* in one? Considering that we have more or less 50 possible such *signiferi* attested at one time, this would argue that the legion was pretty much in its garrison, only nine *centuriae* (about 15% of the total) missing, probably away in a vexillation.

I chose to include in the lists only the ones that are known by name, therefore sixteen, of which one may be a *discens*, plus the nameless *optio signiferorum* not known. Along with these, other five *optiones* and three *tesserarii* tag along.

The three remaining *signiferi* are mentioned in three different places. No. **50**, a veteran former *signifer*, decurion of the Napoca colony, dedicates a votive altar to the goddess Fortuna *Augusta*, for his and his family's health; the dedication is dated September 1st, AD 204. No. **51**, a *signifer* of the legion, originating from Edessa, Macedonia, lived for 40 years and served in the army 20 years. By testament he leaves in the charge of the monument an inheritor, at the indications of two comrades, an *architectus* and a *coriarus*.

Of the five *optiones*, four (nos. **44-47**) are mentioned on the list of veterans CIL III 7688, from Potaissa. No. **48**, *optio legionis V Macedonicae piae fidelis*, is mentioned twice in inscriptions on *pondera*, at the beck and call of his three masters, Severus and his sons.

A total of 12 *duplicarii* are mentioned in the list of veterans from Potaissa; they probably belong to the legionary cavalry or occupy a rank whose signification remains so far unknown. Sharing the same monument are two *equites legionis*, nos. **117** and **118**, the only ones to be mentioned from *V Macedonica.*

[216] Piso 1993, 284.
[217] Bărbulescu 2012, 65-89 = ICLP 6.
[218] Piso 1980, 129.
[219] Bărbulescu 2012, 69.
[220] Bărbulescu 2012, 71 acknowledges this abbreviation, is aware of the existence and status of *optio signiferorum* (page 85), but does not consider this reconstruction.

[221] Bărbulescu 2012, 85. However only *signiferi centuriae*, not *cohortis* not even *vexillarii*. Moreover, the standard bearers are not part of any *officium* of a superior officer.

On this occasion it is worth to be mentioned also another two fragments that attest another *schola*, followed by alt least seven names. This again must be a *schola* of a *coleggium* of *principales*, however their rank remains unknown,[222] for this reason they cannot be integrated in the statistics.

Out of the 14 ***immunes*** belonging to the legion *V Macedonica*, eight are *custodes armorum*, a number of four posts being mentioned besides this.

Out of the seven *custodes armorum*, six are written down on the list of veterans. No. **132** sets up an altar dedicated to the *genio armamentarii*, discovered in the *principia* of the fortress from Potaissa.[223] No. **133** can be attributed to the legion only based on the proximity.

The *mensor* from no. **136** is probably a *mensor agrarius*, and the *architectus* from no. **137** is one of the few to be attested on this very hard to decipher rank.[224] The two veterans, former *immunes*, nos. **138** and **139** can also be *librarii*, a status concealed by their rank as *immunes*.

[222] Bărbulescu 2012, 62-64.
[223] Speidel, Dimitrova 1978; Bărbulescu 2004, 376.
[224] Stoll 2001.

	Name	Rank/Post	Source	Place
		Officium consularis		
1	*Unknown*	*speculator*	CIL III 7688	Potaissa
2	Maximianus	*speculator*	CIL III 7741 = IDR III/5 426	Apulum
3 to 11	9 x *Unknown*	*speculator*	CIL III 7741 = IDR III/5 426	Apulum
12	Iulius Alexander	*beneficiarius*	CIL III 7692	Potaissa
13	Q. Ennius Ianuarius	*beneficiarius*	CIL III 878	Potaissa
14	C. Iulius Maximus	*beneficiarius*	CIL III 826	Samum
15	P. Ael. Marcellinus	*beneficiarius*	AE 1957, 326=ILD 765	Samum
16	P. Aelius Proculinus	*beneficiarius*	CBFIR 525=ILD 769	Samum
17	Scantius Lucius	*beneficiarius*	AE 1957, 327=ILD 771, 772, 773	Samum
18	Oclatius Florentinus	*beneficiarius*	AE 1957, 329=ILD 775	Samum
19	P. Aelius Sextilianus	*beneficiarius*	CBFIR 527=CIL III 7632	Samum
20	M. Aur. Moenenus	*beneficiarius*	CIL III 822	Samum
21	*Unknown*	*beneficiarius*	ILD 778	Samum
22	C. Iul. Melcidianus	*beneficiarius*	CIL III 14216^6=IDR II 41	Drobeta
23	M. Antius Herculanus	*beneficiarius*	IDR II 57	Drobeta
24	C. Iul. Frontonianus	*beneficiarius*	CIL III 987=IDR III/5 21, 36	Apulum
	Collective	*beneficiarii*	CIL III 876	Potaissa
25	Aurelius Vetus	*adiutor off. cornic.*	CIL III 894	Potaissa
26	M. Valerius Alexander	*strator*	AE 1959, 316=IDR II 38	Drobeta
27	C. Maximius Iulianus	*optio praetorii*	CIL III 7765=IDR III/5 243	Apulum
		Officium legati legionis		
28	Iucundius Iuvenalis	*cornicularius*	CIL III 887	Potaissa
29	Aurelius Victor	*beneficiarius*	CIL III 893, 909=ILD 494	Potaissa
30	M. Pompeius Proculus	*beneficiarius*	CIL III 8048=IDR II 644	Sucidava
31	*Unknown*	*beneficiarius*	CIL III 7688	Potaissa
32	P. Recius Primus	*beneficiarius*	AE 1995, 1286=ILD 463	Potaissa
33	Titinius	*beneficiarius*	CBFIR 555=ILD 480	Potaissa
34	Aurelius	*actarius*	CIL III 7706	Potaissa
35	M. Ulpius Firminus	*librarius*	CIL III 7684	Potaissa
36	Aurelius Sedatus	*librarius*	CIL III 885, 7688	Potaissa
37	Petronius Marcianus	*librarius*	CIL III 7688	Potaissa
38	Aurelius Victor	*librarius*	CIL III 909, ILD 494	Potaissa
39	Aurelius Iulius	*beneficiarius*	CIL III 7688	Potaissa
40	C.Valerius Victorinus	*beneficiarius*	CIL III 1584=IDR II 39	Drobeta
		Principales		
41	Aurelius Atellanus	*candidatus*	IDR II 67	Drobeta
42	Septimius Crispus	*candidatus*	CIL III 7688	Potaissa
43	Aurelius Valerianus	*imaginifer*	CIL III 7688	Potaissa
44	Aur. Lucilius	*optio*	CIL III 7688	Potaissa
45 to 47	3 x *Unknown*	*optio*	CIL III 7688	Potaissa
48	Valerius Sabinus	*optio*	ILD 523, CIL III 12645	Potaissa
49	*Unknown*	*optio signiferorum*	CIL III 922, 927, 928, 929=ICLP 6	Potaissa
50	Ulp. Masculinus	*signifer*	CIL III 854=7657	Napoca
51	Q. Philippicus	*signifer*	CIL III 14492=IDR II 203	Sucidava
52	Aur. Marcianus	*signifer*	AE 1967, 398=ILD 504	Potaissa

53	Aurelius	*signifer*	CIL III 922, 927, 928, 929=ICLP 6	Potaissa
54	Iul. Valens/tinus	*signifer*	CIL III 922, 927, 928, 929=ICLP 6	Potaissa
55	Aur. Valens	*signifer*	CIL III 922, 927, 928, 929=ICLP 6	Potaissa
56	Aur. Marcellinus	*signifer*	CIL III 922, 927, 928, 929=ICLP 6	Potaissa
57	Aur. Valens/tinus	*signifer*	CIL III 922, 927, 928, 929=ICLP 6	Potaissa
58	Iulianus	*signifer*	CIL III 922, 927, 928, 929=ICLP 6	Potaissa
59	Quintillianus	*signifer*	CIL III 922, 927, 928, 929=ICLP 6	Potaissa
60	Septimius	*signifer*	CIL III 922, 927, 928, 929=ICLP 6	Potaissa
61	Titus	*signifer*	CIL III 922, 927, 928, 929=ICLP 6	Potaissa
62	Aur. Severus	*signifer*	CIL III 922, 927, 928, 929=ICLP 6	Potaissa
63	Aur. Aurelianus	*signifer*	CIL III 922, 927, 928, 929=ICLP 6	Potaissa
64	Respectus	*signifer*	CIL III 922, 927, 928, 929=ICLP 6	Potaissa
65	Marcellinus	*signifer*	CIL III 922, 927, 928, 929=ICLP 6	Potaissa
66	Maximus	*signifer*	CIL III 922, 927, 928, 929=ICLP 6	Potaissa
67	Herculanus	*signifer*	CIL III 922, 927, 928, 929=ICLP 6	Potaissa
68	Maximus	*discens signiferum*	CIL III 922, 927, 928, 929=ICLP 6	Potaissa
69 to 101	33 *Unknown*	*signifer*	CIL III 922, 927, 928, 929=ICLP 6	Potaissa
102	Ael. Maximianus	*tesserarius*	CIL III 7688	Potaissa
103	Fl. Marcellinus	*tesserarius*	CIL III 7686	Potaissa
104	Aur. C(ri?)to	*tesserarius*	CIL III 935	Salinae
105	Aurelius Statilius	*duplicarius*	CIL III 7688	Potaissa
106	Septimius Alexander	*duplicarius*	CIL III 7688	Potaissa
107	Aurelius Firmus	*duplicarius*	CIL III 7688	Potaissa
108	Aurelius Vitalis	*duplicarius*	CIL III 7688	Potaissa
109	Aurelius Celsus	*duplicarius*	CIL III 7688	Potaissa
110	Aurelius Lu[?cius]	*duplicarius*	CIL III 7688	Potaissa
111 to 116	6 x *Unknown*	*duplicarius*	CIL III 7688	Potaissa
117	Aurelius Victorinus	*eques*	CIL III 7688	Potaissa
118	Aurelius Verus	*eques*	CIL III 7688	Potaissa
119	Vitalis	*principalis*	ICLP 5	Potaissa
120	Das(sius?)	*principalis*	ICLP 5	Potaissa
121	Iulianus	*principalis*	ICLP 5	Potaissa
122	Antiochus	*principalis*	ICLP 5	Potaissa
123	Valentinus	*principalis*	ICLP 5	Potaissa
124	Valens	*principalis*	ICLP 5	Potaissa
125	Quintillianus	*principalis*	ICLP 5	Potaissa
		Immunes		
126	Aurelius Mucianus	*custos armorum*	CIL III 7688	Potaissa
127	Aurelius Maximus	*custos armorum*	CIL III 7688	Potaissa
128	Aurelius Lucius	*custos armorum*	CIL III 7688	Potaissa
129	Aurelius Agrippa	*custos armorum*	CIL III 7688	Potaissa
130	Aelius Valerius	*custos armorum*	CIL III 7688	Potaissa
131	*Unknown*	*custos armorum*	CIL III 7688	Potaissa
132	Aelius Vitalis	*custos armorum*	ILD 503	Potaissa
133	Firminius Bellicus	*custos armorum*	CIL III 7652	Suceagu
134	Aurelius Valentinus	*tubicen*	CIL III 7688	Potaissa
135	Aelius Primus	*tubicen*	CIL III 906	Potaissa

136	Aurelius Castor	*mensor*	ILD 474	Potaissa
137	Aurelius Maximus	*architectus*	CIL III 7688	Potaissa
138	Aurelius Flavianus	*immunis*	CIL III 7688	Potaissa
139	Aurelius Domnio	*immunis*	CIL III 7688	Potaissa

Table I.2. *Immunes* and *principales* of the legion *V Macedonica* in Dacia

Other legions. A number of at least five other legions left evidence in Dacia concerning the ranks of their soldiers. Most of the mentioned petty officers do not write down the legion they come from and this is sometimes very difficult to assume.

The legion that is most active from this point of view seems to be ***VII Claudia***. The Roman fortress of Viminacium gave it a very important role in all conflicts between Dacians and Romans. Its tile stamps can be found in several locations north of the Danube: Drobeta, Pojejena, Gornea, Vršac, Banatska Palanka, Kovin. It obviously took part in the organization of the new north-Danubian province during the time of Trajan, after which it is no longer mentioned in Dacia until the time of Philippus. A series of inscriptions connected with this legion mention that a vexillation was sent north of the Danube to take part in the war against the *Carpi*.[225] It was probably stationed near Romula, the bulk of the inscriptions being discovered here.

There are eight petty officers from this legion who are found in Dacia; the fact that some of them are from the legion *VII Claudia* can be deduced from the location of the discovery of the monument. Among them, three are part of an *officium consularis*, probably not the one of the consular of the three Dacian provinces, one is from the *officium* of the legion's tribune, three are *principales* and one is *medicus*.

No. **1** is *speculator legionis VII Claudiae [[Philippianae?]]*,[226] sets up an altar for Diana, Mercurius *Gubernator* and for the *genius stationis*. We cannot say for sure whether there existed a *statio* at Cioroiu Nou but the phrasing in the inscription is weird because it is a known fact that *speculatores* are part of the *officium consularis* and are missing from that of the legate of the legion. Still, he mentions the legion he comes from, where he remains and where he will return after finishing the special mission he was entrusted with.[227]

A punctured inscription on a bronze button found in Copăceni-Praetorium, Vâlcea County, partially reads *P. A(elius?) C?, b(ene)f(iciarius)*, nr. **3**. The lack of information prevents us from deducing more. The variant of the authors of CBFIR for the name of the character in question is Pacatianus or Pacius/Pacuvianus, the inscription dating from the 3rd century.[228]
On a brick fragment from Romula we find a *CIN? b(ene)f(iciarius)*, nr. **2**. Because of the location of the discovery we assumed that he could belong in the ranks of the legion *VII Claudia*. The authors of CBFIR again assume *Cin[-cius, -ius, -amus]*, and date the inscription in the 3rd century.[229] There is another beneficiary who is attested this time as part of the *officium tribuni* and his enlistment in the *VII Claudia* is uncertain.

From the *principales,* the one who seems to be the highest ranked in an *evocatus*, no. **8**. He dedicates a votive altar to the goddess Nemesis *victrix*. I assumed he was part of a legion, especially the *VII Claudia*, because of the *evocation* procedure employed most often in legions and cohorts from Rome, having the purpose to recall a soldier for military service after his honorable discharge in order to promote him to centurion or to use him in different areas of expertise. While waiting for a centurion posting these *evocati* occupy the first position in the list of *principales*, with the benefits of special payment, sometimes believed to be the equivalent of that of a centurion, and even special decorations.[230]

A fragment from a slab of limestone mentions two high-ranking *principales*, from the tactical ranks, as part of the *centuria IX hastati prioris* of the legion *VII Claudia*, one *signifer* and one *optio* (nos. **9** and **10**). The authors of IDR claim that the two *principales* led a squadron of soldiers from this *centuria* who participated in the building of Romula's enclosure, in AD 248. In fact, they could very well represent the *centuria* in a vexillation or in any other case when the centurion was absent.

No. **12** is *medicus* in the legion *VII Claudia* who received the ornaments of a decurion from the order of decurions from Drobeta, and deceased at 23 years of age. This exceptional character confirms the most recent and plausible theories regarding *medici legionis*, more precisely those indicating that they enjoyed a special status in the army, differentiated from both soldiers and centurions, enjoying increased payment and having a contract with the army for a set period of time.[231]

In the central-eastern part of the province, at Crăciunel, Harghita County, there is a beneficiary of the consular, no. **4**, who, along with his father Aelius Valerianus, dedicates a votive altar to an unknown divinity (whom the authors of the IDR presume to be Iupiter *optimus maximus*). The presence of the two in this location and the setting up of this altar might indicate a *statio* or be simply a detachment of a soldier of the governor staff.

Two other members of the *officium consularis* are attested from other legions. We refer to two *frumentarii*, one in Drobeta, the other one in Sarmizegetusa.

No. **5**, centurion in the legion *IIII Flavia*, former *frumentarius*?, believed to come from Colonia Victrix Camulodunum, Britannia,[232] dies at 56 years of age, his freedman, C. Titius Epipodus having set up the monument. The inscription presents an uncertainty, that of the restoration of the abbreviation *FR*, that could very well belong to the epithet of the legion. The character is

[225] Piso 2000, 218.
[226] Petolescu, ILD.
[227] See also Cupcea 2008, 269.
[228] Schallmayer, CBFIR 538.
[229] Schallmayer, CBFIR 543.
[230] Maxfield 1981, 210-213.
[231] For the whole discussion see Davies 1969 and 1970, more recently Willmans 1987 and Stoll 2004.
[232] More likely he seems to be from Philippi.

an important officer, who took part in the organization of Dacia as soon as the province was created, with the soldiers from the legion *IIII Flavia*, who might have been a previous *frumentarius*. Still, because of the area and of the abbreviations the possibility *centurio regionarius* is just as likely.[233]

No. **6**, a legionary in the *VI Victrix*, from Britannia, *frumentarius*, dies at the age of 19 and appears on a collective epitaph, along with Antonia Bonosa (38 years), Aurelius Constantius (45 years), Roman knight, decurion in Viminacium and P. Antonius (34 years), tribune of the cohort *III Delmatarum*. It is their father, a certain P. Antonius Super, who seems to be the one who has set up the monument. We are dealing with a family of immigrants with a strong military tradition and from its ranks a generation of Roman knights arose.

Another legion represented in Dacia by one petty officer is *XV Apollinaris*. After taking part in the conquest of Dacia it will move from Carnuntum, remaining in the East. No. **11** is a veteran, former *signifer* of the legion *XV Apollinaris*, decurion of the colony. It is likely he took part in the conquest of Dacia and remained with the first wave of colonists, most of the legionary veterans who were the base for the society from Sarmizegetusa and the province.[234] His inheritor, C. Iulius Macer, is a centurion of *XIII Gemina*.

We have left for last a number of six characters whose affiliation must be with one of the legions from the north-Danubian province. In the north, in *stationes* established in several points, there are four *beneficiarii consularis* who do not mention their legion of origin, but for sure it is one of the two legions from Dacia.

Only two *beneficiarii consularis* are found in a *statio* from Porolissum, nos. **15** and **16**. Another, no. **13**, seems to attest something similar, a *statio* on the *limes* and at Buciumi, or to be connected to Porolissum, in very close vicinity.

Another special situation is that of the two beneficiaries of the procurator *aurariarum* from Zlatna-Ampelum.[235] Although the unit where they are enlisted is not mentioned, most likely it is the legion *XIII Gemina*.

[233] See also Speidel 1984, Cupcea 2009, 310-311 and Cupcea 2013.
[234] A detailed discussion about the veteran colonists from Sarmizegetusa in Cupcea 2012a.
[235] For the *beneficiarii procuratoris*, see Dise 1996. For the situation from Ampelum, Hirt 2010, 159-161,165, 259.

	Name	Rank	Legion	Source	Place
1	A[elius?] Germanus	*speculator*	*VII Claudia*	AE 1959, 330=IDR II 141	Cioroiu Nou
2	*Unknown*	*beneficiarius consularis*	*VII Claudia*	AE 1969-70, 556=IDR II 394	Romula
3	P. Aelius ?	*beneficiarius consularis*	*VII Claudia*	CIL III 13797=IDR II 591	Praetorium
4	Valerianus	*beneficiarius consularis*	?	CIL III 7719=IDR III/4 256	Crăciunel
5	C. Titius Ianuarius	*centurio frum/reg ?*	*IIII Flavia*	AE 1959, 314=IDR II 35	Drobeta
6	P. Ant. Victor	*frumentarius*	*VI Victrix*	CIL III 1474=IDR III/2 379	Sarmizegetusa
7	M. Fl. Epstela	*beneficiarius tribuni*	*VII Claudia*	AE 1998, 1110=ILD 129	Urzica
8	Iulius Caninus	*evocatus*	*VII Claudia*	CIL III 8028=IDR II 336	Romula
9	P. Ael. Victorinus	*optio*	*VII Claudia*	CIL III 8047=IDR II 328	Romula
11	Q. Manlius Verus	*signifer*	*XV Apollinaris*	CIL III 1478=IDR III/2 428	Sarmizegetusa
10	Aurelius Probus	*signifer*	*VII Claudia*	CIL III 8047=IDR II 328	Romula
12	M. Val. Longinus	*medicus*	*VII Claudia*	CIL III 14216[9]=IDR II 42	Drobeta

	Name	Rank	Source	Place
13	P. Iulius Firminus	*beneficiarius consularis*	CIL III 7645=CBFIR 524	Buciumi
14	Cassius Erotianus	*beneficiarius consularis*	CIL III 825	Samum
15	T. Flavius Valentinus	*beneficiarius consularis*	ILD 687	Porolissum
16	Cassius Martialis	*beneficiarius consularis*	ILD 701	Porolissum
17	Satrius Felix	*beneficiarius procuratoris*	CIL III 7833=IDR III/3 300	Ampelum
18	T. Aurelius Diocles	*beneficiarius procuratoris*	CIL III 1295=IDR III/3 310	Ampelum

Table I.3-4. *Immunes* and *Principales* of other/unknown legions in Dacia.

I.2.4. Particularities of legionary *immunes*, *principales* and military *officia* from Dacia

As argued in the previous pages of this paper most of those who are sent to the *officia* of the governors come from the legionnaires in the province. In fact, except the *singulares*, all others had to come from legions. This is one of the greatest privileges that a legionary can enjoy, following his efforts to promote towards the ideal, the rank of centurion. Dacia is no exception to the rule. From the moment it becomes a consular province, *officium consularis* has need of a lot of bureaucrats, for all kinds of activities, whom it recruits from the legions in the province. Because most of the evidence dates past the half of the 1st century we can assume that all members from this staff are part from the army in the province, the ones mentioned as being from other legions not being included in the bureaucratic apparatus from Apulum.

Officium consularis III Daciarum is very well documented. It fulfills all the requirements of an *officium consularis* found in other provinces[236] and respects the same rules of organization and functioning. It is attested by a number of 107 persons, 86 of whom are *principales*, 7 guards and 14 *immunes*.[237]

The reconstruction of the upper part of an *officium* of a governor of a province with one legion might appear as it follows: two *cornicularii*, two *commentarienses*, 10 *speculatores*, 60 *beneficiarii*, 10 *quaestionarii*. The numbers indicate a ratio of one *speculator* and one *quaestor* per legionary cohort and one *beneficiarius* per *centuria*, seemingly quite logical.[238] If this recreation is correct then we must add to this group of 84 individuals one *princeps praetorii*, his *adiutores*, *frumentarii* and the multitude of secretaries. The case of the *frumentarii* is a special one; in the *castra peregrina* there fit about 400 persons, so about 12-13 from each legion in the Empire, but they are not the only ones garrisoning that fort. The secretaries must have been very numerous, bringing the total to about 100-150 people.[239] By applying this on the situation in Dacia the principle might give a number of approximately 200-300 people for the staff of the governor leading a two legions province. To these numbers we add the *stratores*, approximately 200 of them, with their own officers and the *singulares*.

The consular governor of Dacia has an attested number of three *cornicularii* as well as one of their collectives (possibly a *collegium*). They are all found at Apulum, which seems only logical, two of them on the building plaque of a *schola speculatorum*, CIL III 14479 = IDR III/5 426. On this list, according to IDR III/5, appear, in order of their ranks, all former or actual *speculatores*. The list begins with three centurions, goes on to mention two *cornicularii* and five possible former *speculatores*, the remaining two columns containing the then serving 20 *speculatores*. This fact might seem to indicate a common enough practice, that of promoting in the same *officium*, in the light that this fact is very rarely documented. Once appointed on an upper post in the *officium consularis* a soldier might hope for a direct promotion to centurion. This however is not valid for the *speculatores*, because in the Empire we know promotions of those serving in this post to *commentariensis* (three examples) or *cornicularius* (one example). This fact might seem strange given that in the whole of the Empire there are at least two promotions from *beneficiarius consularis* directly to centurion in conditions where it is assumed that it is a lower-ranking post than that of the *speculator*, a fact proven even by such a promotion. Still, there are no known promotions from *speculator* to centurion.[240] It is possible that this inscription might be one of the few counterexamples of this theory. Thus, if the three legionary centurions, one from *XIII Gemina* and two from *V Macedonica*, had received this rank promoting from another post, they would have been mentioned in a *collegium* belonging to that post and not in between *speculatores*. According to this principle we would have three new examples of such promotions, at least two examples of promotions towards *cornicularius* and an uncertain number of examples of promotions towards *comentarienses*, all in the same *officium*.[241]

Coming back to *cornicularii*, a province with two legions should pass on to the *officium consularis* a number of three or four *cornicularii*, perhaps two from each legion. The origin of the consular's *cornicularii* found in Dacia is unknown.

Logically, the numbers should be the same for the *commentarienses*. Unfortunately no such soldier-clerk is even found in Dacia, but the presence of their collective in the *officium consularis* is proven by the mentioning of their association next to that of the *cornicularii* and *speculatores*, on a fragment from an honorific inscription (CIL III 7794b = IDR III/5 435). Plus, if the building plaque discussed above would contain evidence about them, they would be the only ones from the north-Danubian province.

Speculatores are very well attested in Dacia, mostly due to the already mentioned construction plate. This monument contains, on two columns, ten *speculatores* on each one, from each of the two Dacian legions,[242] the names being mentioned only in three cases. Besides them we know the collective from the inscription next to the *cornicularii* and *commentarienses* as well as four other individuals. From this group one is mentioned at Potaissa, as enlisted in the legion and the other three at Apulum.

---

236 Rankov 1999, 23.

237 The first centralizing study about the *officium consularis* from Dacia is Dobo 1978, 57-64.

238 Rankov 1999, 23-24; *contra* Ott 1995 and Nelis-Clément 2000, opting for 30 *beneficiarii* from a legion.

239 Austin, Rankov 1995, 152.

240 Breeze 1974, 268.

241 Detailed discussion in Cupcea 2008.

242 With this occasion the *schola speculatorum* is also mentioned, approximately at the same time it is mentioned at Aquincum (CIL III 3524.). Dobo 1978, 58.

The great majority of *principales* from the *officium consularis* is established by the high number of *beneficiarii*, 49. No less than 32 from them originate in the legion *XIII Gemina*. More than half of them, 18, are attested at Apulum. This does not indicate that a great number of beneficiaries worked in the governor's *praetorium*, especially because most of them are attested as veterans or simply passing by. Still, part of this officer group worked as *domi curatores* attached to the governor's residence and certainly helped him with tasks from the provincial 'capital'.
In order to establish as clearly as possible the problem of beneficiaries and of their activities in Dacia, I consider it necessary to make an excursion into the generalities of this subject. During the Republic governors were in the habit of taking with them in the provinces *amici*, members of the family, to assist and counsel them. The first signs of patronage – *beneficium*, date back from the times of the Civil Wars.[243] The earliest mentions of the term *beneficiarius* of a governor date back to the time of Caligula, with the passing of the legion *III Augusta* from the command of the proconsul of *Africa* to its own *legatus*. Thus, Tacitus[244] says that the two magistrates shared their *numerus beneficiorum*. Inscriptions of the governors' beneficiaries appear shortly after, the first probably dating back in between AD 52-60,[245] a *beneficiarius procuratoris Judaeae*. Two others are dated in the 1st century.[246] No *stationes* are known at this time, thus they probably activated at the governor's headquarters. Two examples of beneficiaries who were chosen by two consecutive governors can also be dated now.[247]

Two inscriptions from Lambaesis[248] indicate a number of 50/55 *officiales* of the governor of Numidia, 30 of whom are *beneficiarii consularis*. In provinces with one legion it is likely that their number is double – 60, even if this figure is not certain. Because most references were discovered in *stationes*, it was concluded that the serving period for such a *beneficiarus* was six months, after which he was moved. There are however *beneficiari* who activated at the governor's headquarters, where they assist him with different tasks. They were trustworthy people attached to the headquarters – *domi curatores*, and probably those from the *schola beneficiariorum* are chosen from among them. It is unknown how many of them leave and how many remain, probably in between *stationes* they serve for a while in the *praetorium*, for instruction and report. They are associated with other ranks in those *scholae* with which they probably had professional connections: *cornicularii, speculatores, quaestionarii, frumentarii*, centurions. They get together with *speculatores* and *quaestionarii* in legal tasks, involving arrest and interrogation. They are even associated with *immunes*(=*librarii*?), probably for administrative tasks, or *stratores* probably as responsible with the stables.[249]

Papyri present them as carrying out courier missions, financial auditors, inspectors on behalf of the ruling authority or as agents of law and order[250] In many cases they are tasked with forwarding the governor's orders, making sure they are followed and reporting back. The difficulty in promoting beneficiaries lies in the small number of higher-ranking posts available in an *officium consularis*. In provinces where there are no legions, recruitment is done from the armies of the neighboring provinces: for Gallia from the Germaniae, for Dalmatia from the two Moesiae and the two Pannoniae. The service record of a beneficiary is limited, after which he returns to his troop.[251] They are not found in senatorial provinces, except for the particular situation of Africa.[252]

Trajan was the one who practically initiated a new status for beneficiaries. He decided the creation of *stationes* populated with beneficiaries of provincial governors. Most of them appeared between AD 110-120 in the whole Empire, in different provinces and populated in the same way. All these factors indicate a central, imperial decision. Epigraphic evidence regarding these beneficiaries differs completely from those from the 1st century. Firstly, they are not connected to funerary monuments but to votive altars, most of them dedicated to Iupiter *optimus maximus*, set up while serving in the *statio*. They indicate that those serving in *stationes* pay a greater importance to their mission than those serving in the governor's headquarters, where we find only one altar for a period of 50 years.[253] Two *stationes* definitely created by Trajan are those from Sirmium[254] (from 80 altars 4-5 can be dated before AD 150) and Celeia, Noricum (the first altars are dated AD 110, 115). The two are both ancient cities, of the 1st century, both laying at the end of a major road in the Empire, coming from northern Italy, past the Alps, along the valleys of the rivers Sava and Drava and finally reaching the Danube. Thus both *stationes* have a strategic position considering Trajan's plans. Celeia is active until AD 160, and then is reused between AD 190 and 217, while Sirmium is active without interruption until AD 230. Other isolated altar finds from this period are at Virunum and Iuvavum (Noricum), Sebastopolis (Cappadocia), Charax (*Regnum Bosporanum*). They come to support the theory that the two examples mentioned above are not an isolated

243 Caesar, *BC* 1.75, 3.88. Nelis-Clément 2000, 59-61.
244 *Hist.* 4.48.
245 AE 1967, 525.
246 CIL XII 2602 (79) and CIL VIII 27854 (81-83).
247 Dise 1997, 274-275.
248 CBFIR 783 and 784.

249 Ott 1995, 73-76; Nelis-Clément 2000, 113-126.
250 Rankov 1994, 227-229.
251 In provinces where the service period can be dated this is six months except for the beneficiaries of *vigiles* from Ostia, who serve for only four months. Ott 1995, 105.
252 Ott 1995, 77-78, 82-84.
253 CIL III 14362, Virunum.
254 For the complete discussion about the *statio* from Sirmium, see Mirković 1994, 348, 357: From Trajan until AD 157 – 4 altars, between AD 157-185 – 17 (1 dated) altars, between AD 185-202 – 26 (8 dated) altars, between AD 203-209 – 8 (7 dated) altars, between AD 221-231 – 12 (8 dated) altars. Promotions are mentioned: from *frumentarius* to *beneficiarius consularis*, and from *agrimensor legionis?* to *speculator*. Confirms service period of six months and repetition of the mission in the same *statio*. Also Nelis-Clément 2000, 141-148.

phenomenon but rather a generalized one. The altars are dated around AD 120 and the lack of later ones might indicate the dismantlement of these *stationes* around AD 130.[255] At this moment appointing such a clerk was indeed a favor, a *beneficium*. The correspondence between Pliny and Trajan is proof of this.[256]

Following, Hadrian was completely uninterested by this matter because by AD 130 only the *stationes* from Sirmium and Celeia were still in use and in between AD 125-150 there is no evidence for any other. Antoninus Pius and Marcus Aurelius compensated for Hadrian's lack of interest, giving provincial administration and the governor's staff a new impulse. An imperial initiative is again involved, because it is applied in a very short period of time in the whole Empire. After AD 150, epigraphic evidence begins to surface in other points as well: Rigomagus (Germania Inferior), Pontailler-sur-Saône (Germania Superior), Praetorium Latobicorum[257] (Pannonia Superior), Montana and Histria (Moesia Inferior), Tyana (Cappadocia). This increase in the number of *stationes* is part of the operational axis of imperial actions to this goal. Around AD 150 there is a great increase in the activity of the points from the already mentioned road: seven altars at Celeia, several at Sirmium (dated post 157). All this is until around AD 160 when Celeia closes, probably replaced with Praetorium Latobicorum and the altars from Sirmium are missing. Perhaps due to the Marcomannic wars the number of altars dated between AD 161-170 drops to seven, in six provinces.[258]

From this point on it will follow an explosion regarding the activities of these *beneficiarii* in *stationes*. Some are isolated finds, evidence of a short activity in that place[259] but five of them have great discoveries: Montana – mining area – five altars, two in Germania Inferior – Nettersheim – seven altars, until AD 227, Rigomagus – seven altars, until AD 242, three in the *agri decumates*, Germania Superior[260] – Stockstadt – 21 altars until AD 208, Jagsthausen – five altars, until AD 186, Obernburg am Main – 75 altars until AD 224.[261] The increase in the number of altars means an increase in the number of *stationes* and not necessarily that of the beneficiaries. The organizational reform initiated by the Antonine dinasty results in the transformation of the relation between the governor and his beneficiary from a personal one to an institutionalized one, organized in the confines of an *officium consularis*. The term *officium consularis* appears for the first time connected with a *beneficiarius* on a monument from Tyana, Cappadocia, around AD 150,[262] the term being transliterated in Greek: *ἐξ όφικίου*. The first *beneficiarius consularis* is found at Montana, in AD 155[263] and later on the formula will proliferate, so that after AD 170 we have only four beneficiaries that can be attributed to a certain governor. This reform includes service length and recruitment manner. There is evidence for length of service greater than the consular's time as governor and for the 3rd century we have evidence that their service was apparently permanent, unless promoted. From this moment on *beneficiarius consularis* becomes a *principalis, duplicarius*. These movements shift the loyalty of the men from the governor's person to the governor as an institution. Also now the post is entered into the lower legionary hierarchy, chances for promoting to centurion rising.[264]

The word *statio* has multiple meanings: from guard post or meeting place to military installation with general purpose. In connection with *beneficiarii* it means headquarters but also simply mission.[265] The terminology used for active service was *agens curam stationis* or *habet stationem*, the end of service, usually six months in length is expressed by *expleta statione*, *emerita statione* or *exacta statione* and repetition of the delegation in a *statio* by *iterato statione*. Even the missions of the beneficiaries of procurators *metallorum* in a mining area are called *stationes*. Those sent in neighboring provinces remain members in their own unit but, temporarily, they are individual embodiments of the imperial administration, subordinated to the authority of the province where they dispatch. But such a mission is not called *statio* because this term seems to define a well-established spatial and temporal notion; outside the

[255] Dise 1997, 275-277. Relation between them and the governors who selected them was just as personal as before, as proven by the phrasing *beneficiarius + governor's name + governor's rank*. The service period of the beneficiary was limited in time, after that the soldier returning to the unit of origin.

[256] *Ep.* 20, 21, 22, 27.

[257] For the *beneficiarii consularis* at Praetorium Latobicorum, see Sasel Kos 1995.

[258] Dise 1997, 279-280.

[259] For the *beneficiarii* in Aquileia, see Buora 1994.

[260] A detailed discussion about the *beneficiarii* from Germania Superior in Schallmayer 1994, 189-191. Many places where altars containing beneficiaries share the same topographical placement, near forts. Those closest to roads and water sources have a greater role, just like those nearby mineral resources or connected with outside trade and the traffic of persons or goods through the *limes*. The beneficiaries of the governor of Germania Superior organized themselves in a petty officers corps with a highly developed self-awareness, similar associations for beneficiaries being unknown. A single auxiliary beneficiary is known here, probably with economic duties, just like the rest stationed in *stationes*. All altars are dated between AD 150-240, probably the province experienced great economic developments in this period. *Beneficiarii* active here come from the two legions and one cannot establish a territory set aside for each legion from this point of view. Their activities must be considered part of the economic and financial control of the province, police duties coming second. Also, Nelis-Clément 2000, 133-161.

[261] Details about this recently discovered *statio* and its sacred area in Steidl 2005.

[262] IGRR III, 6800.

[263] CIL III 7449.

[264] Dise 1997, 280-283. The Antonine dynasty modified the length and mandate of such a *beneficiarius* in stationes. If in between AD 110-150 at Celeia there is one altar for each procurator, in the period after AD 150 we have two beneficiaries for each procurator. This does not indicate a double number of clerks in a *statio* but the division of the length of service. It is a phenomenon that can be observed only here. Under Marcus Aurelius and later length of service in *stationes* varies for each province, sometimes every decade. Up until the 3rd century governors leave these details for the chiefs of staff from their *officia* to handle but they can decide the opening or closing of *stationes* according to the needs of the moment. For a detailed discussion about length of service and the way beneficiaries manifested themselves in *stationes* see also: Dise 1997a, Schallmayer 1994, Isac 1994.

[265] Dise 1995, *passim*; Fuhrmann 2012, 205.

officium the petty officer is a member. In inscriptions it could also refer to mailing service. When considering spatial distribution we notice that the placement of *stationes* is a general one, at the level of the whole Empire, with four categories:[266] A – in cities; B – on the *limes*; C – on important roads, crossroads or ports; D – in mining areas or other imperial enterprises.[267]

One notices an important quantitative domination of the C-type all over the Empire. In the northwestern provinces type C dominates and type B is also present on the Rhine forntier. The few A-types that exist can be explained away because of the lack of native urban culture in the area. In the Danubian provinces type A is more numerous, type C are fewer, type B is found only on the Danubian *limes* and type D is also present. In the Eastern provinces type A clearly dominates because imperial administration was based here on already existing urban structures. Egypt is special as *stationes* are placed in the administrative centers of the *nomoi*. In northern Africa most of them are type A, very few type B. In Hispania almost all are type D. When analyzing geographical distribution one notices than the density is much greater in European provinces than in the Eastern ones.[268]

Papyri bring proof than in Egypt there were a lot of *stationes* along the Nile, starting with AD 117-119. The *beneficiarii* serving here have police duties, alongside the decurions or the centurions. Such service periods in *stationes* are found all across the Empire, starting with Trajan. Their function is not clear, probably they formed together a conglomerate but one notices they are massed in areas exploited under imperial monopoly. Many of them can be found on the *limes* but they cannot be considered customs since this was an internal matter, not to be collected when coming from the *barbaricum*. Their presence along the main roads, at important crossroads or around Roman camps might indicate military tasks. A plausible explanation for their presence in these areas is their possible role as intermediaries between the governor and his army. With this in mind, after finishing their service that might last only six months in such a *statio*, they probably return to the governor's headquarters, giving him a detailed rapport about the real situation in the field and the military installation in the supervised area.[269]

The earliest *stationes* are those belonging to type A, half of them being already mentioned before the Severan dynasty, at the earliest under Trajan/Hadrian. Those on the *limes* appear only after the half of the 2nd century, earlier evidence being quite scarce. Type C is very heterogeneous from this point of view and type D begins to appear during the Marcomannic wars.[270]

Although in inscriptions dedicated by these beneficiaries appear formulas *pro salute sua et suorum* they refer to the family because, as the *papyri* inform us, the personnel of a *statio* is made up of only one *beneficiarius*. There is also a type of *statio* with two *beneficiarii* attested at the same time, however they are very rare. Still, some epigraphic evidence might indicate that sometimes the beneficiary was assisted in his tasks by a number of clerks or soldiers (*exceptores*) thus setting up the dedications together. Probably they collaborated with all sorts of soldiers, depending on the nature of their mission. In all other cases no evidence regarding a command belonging to beneficiaries can be found.[271]

We have a wide variety of attested local policing tasks in the case of beneficiaries. They are considered agents of law and order both by the provincial administration and the peregrine inhabitants. These are closely connected with administrative tasks and enforcement of justice in the provinces where *beneficiarii* are agents tasked with detaining and supervising convicts.[272]

The beneficiary also had an active economic role. This means he is the delegate of the authority for collecting taxes and applying financial provisions, as well as preventing any kind of tax evasion. Also the placement of some *stationes* in key transit points for merchandise, in ports or at important crossroads indicates their ability to collect taxes and to supervise the system of *cursus publicus*.[273] The presence of consular beneficiaries in mining areas or in other imperial enterprises cannot be easily explained. Their tasks probably varied: insuring order, collecting some taxes or even serving as liaison between the mining procurator and the governor.[274]

These *officiales* are spies who prevent enemy movements, representatives in the field of the executive power, play a role in prosecutions, in proving facts, in trials, they arrest, interrogate and submit documentation to the judges. The population refers to them for problems, since they are agents of the authority; they send briefs at the center and play an important role in applying the penal code. Their involvement in financial life was due to their role as investigators, auditors, verifying the legitimacy and correctitude of entrepreneurs. Finally, they supervise all mines owned by the state. All these indicate that they had to be properly trained to read and write, in mathematics and law, as seen in the *papyri* or different representations.[275]

A rather prudent conclusion that can be drawn from this variety is that the *beneficiarii consularis* held as many duties as the governor would ask them to. In fact, they are the agents of the governor in the most distant, peripheric places of the Empire, in charge with whatever task that they are entrusted. The situation of Egypt cannot be

[266] Ott 1995, 85-87; Nelis-Clément 2000, 175-179.

[267] Ott 1995, 88.

[268] Ott 1995, 100-101. The separate study of *stationes* on each frontier of the Empire, in Nelis-Clément 2000, 161-173.

[269] Rankov 1999, 27-29.

[270] Ott 1995, 102-103.

[271] Ott 1995, 107-112; Fuhrmann 2012, 205.

[272] Hirschfeld 1891, 15-33; Ott 1995, 115-129; Nelis-Clément 2000, 220-243. Even if the actual police duties belong to the *milites stationarii* and *frumentarii*; see Fuhrmann 2012, 208-220.

[273] Ott 1995, 130-150; Nelis-Clément 2000, 243-259.

[274] Zwicky 1944, 87-88; Ott 1995, 152; Nelis-Clément 2000, 259-264; Hirt 2010, 42-44, 129-130, 199.

[275] Ott 1995a, 286-287.

extrapolated, their role in police duties and supervision of roads and customs being also overrated. These general delegates are connections between the central provincial authority and the field.[276]

The highest rank that a legionary *beneficiarius* could achieve was *beneficiarius consularis*. It can be stated that he had to be an educated man, trained in as many aspects as possible regarding provincial life, economic and military matters in order to ensure a continuous and proper flow of information towards the *officium consularis*. That is why they originate from a multitude of posts or ranks, divisible in three categories: on one hand from the tactical ranks of the legion, people who had tactical command experience but with limited administrative experience. On the other hand most come from posts of beneficiaries of officers/commanders of troops: tribunes, *legati*. They already have the administrative experience accumulated from that post and promote, as a sign of their efficiency, in the highest specialized rank. Last, they have a special connection with the *frumentarii*, several such *frumentarii* being promoted to *beneficiarii consularis*. The joining of the word *candidatus*[277] to their title does not mean that they are waiting to promote to centurion, only the opening of a post in the *officium*. For most of them this seems to be the culmination of their careers, especially since many are retained in service for several years and their missions are repeated – *iterato agens statione* sau *tertia statio*.[278] A small number among them promote to *commentariensis* or *cornicularius*, perhaps *adiutor principis praetorii*. An even smaller percentage might be promoted as *trierarchi* of the fleets or as auxiliary decurions/centurions and after the half of the 2nd century directly to legionary centurion,[279] where they can advance as high as *primi ordines*. Towards the half of the 3rd century some might cross the social barrier, being attested as *petitores militiae*.[280] This rank can still be found under Diocletian and *protector praefecti legionis* seems to have evolved from it.[281]

The problem of beneficiaries and that of *stationes* in Dacia has been for a very long time on the agenda of Romanian but also foreign researchers.[282] While as far as Romanian academics are concerned, some patriotic passions might have influenced historical interpretations,[283] the foreign ones have integrated the corps of beneficiaries from Dacia in the whole system of the Empire also noticing that Dacia is in no way different from this point of view.[284] More recently J. Ott and Jocelyne Nelis-Clément gathered evidence in their monographs dedicated to this type of soldier.

In two papers from 1995, J. Ott gathers all evidence regarding *beneficiarii consularis* from Dacia and considers that each epigraphic discovery must be a *statio*. Thus, all types of *stationes* identified by him in the Empire are also attested in Dacia as follows: A – Napoca, Porolissum and Sarmizegetusa; B – Buciumi, Samum and Crăciunel; C – Praetorium and Micia; D – Alburnus Maior and Ampelum.[285] Nelis-Clément refers in general to the *stationes* from Dacia, without classifying them absolutely; those from the north are dependent on the northern *limes*, those from the Apuseni Mts. as connected with the gold-mining area and those from Oltenia as controlling the trade with *barbaricum*.[286]

Perhaps these approaches are not very fortunate, especially because of the contents of epigraphic evidence. A number of three beneficiaries are mentioned at Sarmizegetusa, along with one at Apulum, who settles here after discharge. They are not evidence for the existence of a *statio* in Țara Hațegului, because they are mentioned either as veterans settled here or as relatives or related to inhabitants. The situation at Micia is similar, where, based on three beneficiaries found here, he confirms the supposed or partially proven so far building activity of the legion from Apulum, in this obviously important place, that might also offer the opportunity for the creation of a special observation post. Alburnus Maior on the other hand fulfills all conditions for the placement of a *statio*: three mentions of beneficiaries and obviously a crucial economic area where, in order to ensure security and authority, the presence of such officers is needed. Finally, one last beneficiary from the legion *XIII Gemina* is mentioned at Ampelum. His presence here can be explained by his appointment by the governor, separate from the *officium* of the *aurariarum* procurator, who has his own beneficiaries, probably enlisted from among the legionaries as well.

The 13 consular beneficiaries from the legion *V Macedonica* are attested in four places in Dacia. One is mentioned at Apulum, but as a veteran and decurion of the colony. At Potaissa two such individuals and a collective are mentioned, probably this is the place they come from. Finally, two are attested at Drobeta, a traditional place for the recruits of *V Macedonica* legion, them not being the only soldiers belonging to this legion who are attested here, probably also the place for the existence of a *statio* overlooking the bridge over the Danube. Four other consular beneficiaries are attested in northern Dacia, without information regarding the legion they come from, although this must be one of the two

[276] Detailed arguments in Dise 1995, especially 81-83.
[277] CBFIR 784, Lambaesis.
[278] Sometimes even at a distance of four years – Mircović 1994, 349.
[279] Although it was believed this was so starting only with Severus, Domaszewski 1908, 33-34; Speidel 1992a.
[280] Ott 1994, 244-246; Ott 1995, 30-34. Claims all *beneficiarii* are *sesquiplicarii*. For such an example, see Haensch 2001a. For the petitores, see Le Bohec 2012.
[281] Domaszewski 1908, 34.
[282] For a history of Romanian efforts concerning this subject, Isac 1994, 209-211; Ardevan 1991 and1994; Opreanu 1994; Isac 2001, 48-55.
[283] Especially considering *Ans(amenses?)* a tribe of free Dacians etc. V. Pârvan, C. Daicoviciu, I.I. Russu, *apud* Isac 2001, 48-55.
[284] Dobo 1978, 59-62, inventories the *stationes* of beneficiaries from Dacia and their personnel attested epigraphically.
[285] Ott 1995, 94-95. In the article of this same author Ott 1995a, 289, he also adds Drobeta and Romula as places with finds, but doesn't include among the *stationes* and does not give details. Dobo 1978, 59-62, who adds even the legionary fortresses in his list of *stationes*.
[286] Nelis-Clément 2000, 164-166.

stationed in Dacia. One is mentioned at Samum, one at Buciumi and two at Porolissum, perhaps charged with the mission of supervising passage through the customs attested here.

A location where the highest number of beneficiaries sent in the territory is the undoubtable *statio* from Căşei-Samum. From the legion *XIII Gemina* there are at least three petty officers detached here and one at Napoca who might in fact be serving at Căşei, and from the legion *V Macedonica* there are no fewer than eight such men. A lot of ink was spilled on the subject of these beneficiaries and their role in northern Dacia, but none of the papers seems to have settled the problem in a definite manner. All our problems come from the understanding of *REG ANS*, mentioned on several altars set up by consular *beneficiarii* from Căşei-Samum during the reign of Gordian III. This was associated with an indigenous population from northern Dacia that supposedly was controlled by the soldiers from a *statio*, then annexed to the province and transformed in a *regio*,[287] with an area from the imperial domain including the salt mines from Dej,[288] with a commercial tax – *ansarium* or with a meander from the river Someş.[289] Still, a century has passed without any new evidence coming to light and a definite solution has not been found.[290]

Given that the abbreviation cannot be explained in a satisfactory manner we must work with what we have at our disposal. There are at least two different stages in the existence of military installations in the area around the auxiliary fortification from Căşei. The first one dates back to the reign of Severus Alexander and marks the existence of a *statio*, proven by ten of the 14 monuments discovered in the auxiliary fortification.[291] Only in four cases is the existence of a *statio* mentioned concretely, in different ways: *agens in munere stationis*,[292] *agens curam stationis*[293] or *iterato agens statione*.[294] These references make it obvious that such a military installation existed in the area, as well as the fact that the *beneficiarii consularis* were in charge of it, some even repeating this mission. In the case of only one the activity in the *statio* is named *in munere*, perhaps indicating he is not the chief of the *statio* in favour of his colleagues. In the other cases the position of *curam agens* indicates a complete command and coordination of the whole activity connected with that *statio*. At this stage there is no mention regarding the name of the place, nor about any special missions the beneficiaries might be charged with, thus we must include this *statio* in the system of imperial provincial administration and of delegation of the governor's authority in the field and we can consider it rather as a mission, not necessarily as a military building or installation.

A second stage in the existence of a military installation around Căşei dates back to the time of Gordian III. In the four epigraphic monuments dated now the mention regarding any *statio* disappears. This does not mean that the *statio* from Căşei is gone, not even that it was transformed, only that is simply not mentioned. On the other hand there is the mention of a *regio*, named *regio Ans(?)*. Inside this *regio* that is in no way different from other such *regiones* from the Empire, organized in the needs of military control over an area of military interest, and placed under the command of *regionarii* centurions, there also activate *beneficiarii consularis* and this is not true only for Căşei. The nature of their activity here is illustrated by a supplementary mention in the inscriptions, in the form of *agens sub signis Samum cum regione Ans.*,[295] *agens sub signis regione Ans.*[296]or *agens Samo cum regione Ans. sub segnis*.[297] Although no definitive variant for the abbreviation *Ans.* has been found, perhaps the other references can be of use. First of all, *beneficiarii* from this period are no longer *curam agentes*, meaning they no longer lead the post where they are sent. This is somewhat normal because this time we are not dealing with a simple *statio* but with a territorial and administrative unit under military control, a *regio* that, according to the analogies from the Empire, should have been led by a *centurio regionarius*.[298] The beneficiaries are *agentes sub signis*, meaning members of a military detachment that could have a *signum*, thus probably also a *signifer*, and other officers. As to the territorial delimitation of their competence it is also mentioned as *Samum cum regione Ans.*, meaning the fort, *vicus Samum* and a *regio* in the neighboring area. We cannot make definite observations about the nature of this *regio* because the abbreviation *Ans.*, which might indicate its name and may or may not be connected with Samum, does not offer the necessary details. Still, the possibility that an indigenous population was named after the river Samus, *Ansamenses*, and then with the word *regio* written after it, or the eventuality of the extension of Roman rule over a *civitas* transformed in a *regio*, seems to be the least plausible.

The apparent overlapping of a *regio* over the territory of an auxiliary fortification is also strange although the place of discovery of the monuments does not necessarily establish such a fact. To this end, the presence of legionaries here would be pretty strange, in order to control a territory that *de iure* is under the supervision of the auxiliary unit and its commander, though it is not impossible to find such a situation, as noticed in the analogies from Montana.[299] A *regio* is attested there, named after the most important settlement in its territory, *regio Montanensium*, an auxiliary troop being stationed

[287] Previous research, V. Pârvan, C. Daicoviciu, I.I. Russu, M. Macrea, D. Tudor, *apud* Isac 2001, 48-52.

[288] V. Vollmann, *apud* Isac 2001, 53-54.

[289] Opreanu 1994, 71-74.

[290] Isac 2001, 57-58.

[291] CIL III 826, 823, 825, 7632, CBFIR 527, ILD 771, 772, 773, 774, 775. The place of discovery for all the monuments is the auxiliary fortification, Isac 2001, 52-53.

[292] CIL III 825.

[293] ILD 771 and 772.

[294] ILD 775.

[295] CIL III 827=7633.

[296] CIL III 822.

[297] AE 1957, 326.

[298] Although there is no such find in the area, not even that of a simple *centurio legionis*.

[299] Speidel 1984, 185.

next to the city, at least up to a point.[300] Besides this troop a series of *regionarii* are mentioned here; they have policing duties in this district and they are different types of legionary *principales* – *optio leg. XI Claudiae, vexillarius eq. leg. I Italicae* – who mentions his mission here by wordings of the type *agens regione* and are under the command of a *centurio regionarius*, this time attested epigraphically.[301] Later on, after the departure of the auxiliary troop stationed here, these *regionarii* seem to take complete control over *regio Montanensium*, and seem to gather between AD 253-258 in a *numerus collectus regionariorum*, an *ad-hoc* military unit created by joining soldiers from other units sent in the area, ensuring the continuity of the garrison in a region that was obviously very important.[302]

The situation seems to be quite similar with that from Căşei. Here too we find a *regio* that we think was named after the most important settlement from its territory – *Samum*, we find evidence of *regionarii* from among the *principales* from the legions of Dacia – *beneficiarii consularis*, who are explicitly sent here with military tasks and apparently formed in an *ad-hoc* unit, mentioned in the inscriptions found here under the name *sub signis*. It is hard to determine who led these *regionarii*, probably a centurion. As to their role here, we can only guess. By extending our analogy with what we know from Montana, which is not the only case, we might conclude that presumably because an auxiliary unit was absent from the fort in this troubled period for Dacia and the Empire, even though temporary, it was replaced with a *numerus* made up of *regionarii*. Still, such suppositions need to be proven, which is impossible for the time being.

To conclude, consular beneficiaries were very active in the area of Căşei in the 3rd century. The nature of their activities here seems to have changed sometime under Gordian III, the area being organized under the form of a *regio* and placed under military control of *regionarii* designated from the consular governor's *praetorium*. The reason for sending these legionaries in areas where an auxiliary troop is stationed is unknown to us but the new situation from the half of the 3rd century might illustrate a major change in the provincial administration of this area. Unfortunately, these monuments of beneficiaries probably reveal to us the 'tip of the iceberg', the evolution of research being dependent on new discoveries.

The strict classification imposed by some researchers for *stationes*, or other places where mentions of beneficiaries were found, is not totally valid. By ignoring the contents of epigraphs or sometimes the geographical location one might probably make confusions or unfortunate separations. Thus, as far as type-B *stationes* in Dacia, at least two of them might be considered also type C (Samum and Crăciunel) and from those considered type A, two are unlikely (Sarmizegetusa and Napoca) and one might as well belong to types B or C (Porolissum). The only categorization standing up is that of type-D *stationes*, those from mining areas. Thus, we must treat this phenomenon in a more general manner, to the point of correlating evidence with others from the Empire and to attempt to frame them in a unitary system of provincial administration, not necessarily by attributing different tasks to different types of posts. Papyrologic evidence from Egypt and the East have especially shown that the missions of these beneficiaries might differ a lot, ranging from local police duties and traffic control, to courier and information gathering missions or even to maintenance of military installations.[303]

The significant number of attested beneficiaries might indicate the fact that this is a post from which it is very difficult to promote, superior posts being in a very small number. It is probable the rank where most legionary careers ended, especially due to the great number of veterans, former beneficiaries, who are attested.

Ranked right under the beneficiaries in the *officium consularis* are the *quaestionarii*. Their role as judicial and investigative agents is well known, that is why, along with the *commentarienses*, they are present only in the staffs of superior officers who also have civil jurisdiction, the governors. Six such *quaestionarii* are attested at Apulum, all of them mentioned on an honorific or votive monument (CIL III 7803 = IDR III/5 459), on a list.

A single *frumentarius* is attested in the *officium consularis* from Dacia, coming from the soldiers of *XIII Gemina*, at Sarmizegetusa. He is not serving away on duty but he comes from a family from the local elite. Two others are attested in Italy, while serving in *castra peregrina* or in a *statio* in Via Appia.

Six *stratores*, as well as one of their collectives, are attested in Dacia. Five of them, and also the collective, are part of the legion *XIII Gemina* and are mentioned at Apulum. Only one comes from the *V Macedonica* legion and is mentioned at Drobeta, probably sent here on some mission. Another is promoted, becoming centurion of *XIII Gemina*.

Last but not least we must mention an *optio praetorii*, adjutant and assistant of the centurion who effectively ruled the *officium consularis*, *princeps praetorii*. He comes from the legion *V Macedonica* and is attested while being on duty at Apulum. His role and position is obviously superior probably to any other post in this staff and the mentioned person awaits promotion to centurion.

Most *immunes* serving in an *officium consularis* are secretaries. They must be in great numbers in order to serve each and every higher ranked official above them. In Dacia we find 14 such secretaries. Two of them are the direct assistants of the *cornicularii*, as *adiutores*. The five

[300] *Coh.I Claudia Sugambrorum*, until AD 129. Speidel 1984, 185.
[301] Speidel 1984, 185.
[302] Speidel 1984, 188.

[303] Whole chapters regarding the functions of beneficiaries, proving their versatility, in Ott 1995, 82-155; Nelis-Clément 2000, 211-268. See also Dise 1995.

librarii consularis from Dacia come from the legion *XIII Gemina*, attested at Apulum, Ampelum or Gherla. The remaining seven are *exceptores*, secretaries in charge of writing reports. All of them seem to originate from the legion at Apulum, six of them being mentioned on an altar dedicated to Minerva, the protective goddess of all these literate men, only one being mentioned at his probable birthplace, Napoca. The six set up a collective dedication in honor of Minerva (AE 1964, 193 = IDR III/5 263), which might indicate that they are all *exceptores* serving at one time in the *officium consularis*, perhaps indicating that half of them are selected from *V Macedonica*.[304]

The *singulares*, both equestrians and infantrymen, are separate from these *officiales* but they are part of the staff of the provincial governor. The governor's guard, made up of auxiliaries, contains both riders and infantrymen, under the names *equites singulares* and *pedites*. A series of such *singulares* are mentioned in Dacia, apparently all of them originating from auxiliary cavalry units. Still, tile stamps give evidence of a three-way organization of this guard in Dacia: *pedites singulares, equites singulare* and *numerus singularium*.[305]

What does the office of Dacia's governor lack? So far we have no evidence regarding *exacti, notarii, interpretes* or *haruspices* serving Dacia's governor but it is certain that they existed. In any case the most important posts are mentioned, some in great numbers, leading us to believe that the *officium consularis III Daciarum* was a fully functional provincial bureaucratic apparatus.

Other *officia* of superior officers from the legions in Dacia. The staff of the legions' legates is similar to that of the governor, although lacking a series of pieces that in the case of the governor reflect his other attributions beyond military command. The two legions from Dacia have a number of 20 petty officers mentioned in the legates' *officia*. More than half of them are *principales*, the rest being *immunes*, most of them secretaries. We know of only one *cornicularius legati legionis* in Dacia; he is heading the *officium legati legionis V Macedonicae*, at Potaissa.

He is closely followed by 12 *beneficiarii*, the majority, seven, coming from the legion *XIII Gemina*; of these seven only one is mentioned outside Apulum, as veteran, living in Apoldu de Sus. The remaining five come from the *V Macedonica,* four are mentioned at Potaissa, one, more special, mentioned as beneficiary of the legate of this legion, Tiro,[306] probably at Sucidava. One notices that similar to beneficiaries of the consular this is the post most numerously mentioned, being probably along with the other the ending point in the career many legionaries.

The batch of *principales* finally comes to an end with two *actarii*, heading the huge corps of secretaries-*immunes* and supervises the writing of the legion's *acta*. There is one for each legion, mentioned at Apulum, respectively Potaissa.

The corps of *immunes* contains ten names, all *librarii*. The vast number in which they are mentioned indicates they were in great numbers in the legion. It is more than certain that each *centuria* had to have a *librarius* while those recruited inside other *officia* formed a group composed of hundreds of people. Six of them are legionaries from *XIII Gemina*, one at Ampelum and the rest at Apulum, and the four from *V Macedonica* are all attested at Potaissa.

The *officia* belonging to the tribunes of the legions are poorly evidenced in Dacia, with only five persons, four *beneficiarii* and one *secutor*. That *secutor* is some kind of guard of the *tribunus laticlavius* of *XIII Gemina,* but we know little of the *beneficiarii*; they placed between the *immunes*, although close to the tactical ranks. Two of them, from the *XIII Gemina* legion, are mentioned at Apulum and the two from *V Macedonica* are split between Potaissa and Drobeta.

Only the *officium praefecti legionis XIII Geminae* is attested through four characters. One of them, *cornicularius*, was previously *adiutor offici rationum*, thus enjoying a brief career inside this *officium*. Alongside him there we have evidence for two *beneficiarii* and one *librarius*, the last one at Sarmizegetusa.

The *procurator aurariarum* has his headquarters at Ampelum and he supervises and administers the gold mining area from the Apuseni Mountains; however he is not included in the province's military structures. We find in his staff two *beneficiarii*, probably legionaries, recruited from *XIII Gemina*, both of them discovered while serving on duty at the headquarters from Ampelum.

Military ranks are very well evidenced as far as the legions from Dacia are concerned. We have a number of 66 *principales* found on all ranks. Out of them 26 represent the tactical ranks, while the rest stand on both sides. The highest ranks of military *principales* are *candidatus* and *imaginifer*. We know eight such *candidati*, awaiting promotion to centurion, six from the *XIII Gemina* and two from the other legion; one of them is again found at Drobeta. Also we know one *imaginifer* from each one of the two Dacian legions. We must mention that for the legion *V Macedonica* the list of veterans (CIL III 7688), most of them former *principales*; covers this aspect is very well, as far as this legion is concerned.

All three tactical ranks are very well represented in Dacia. 12 *optiones* open the list, seven from the *XIII Gemina* and the other five from Potaissa. They are completed by a horde of *signiferi*, three of them the heads of their associations, meaning *optiones signiferorum*, two from the legion *XIII Gemina* and one of *V Macedonica*. Four *signiferi* also belonging to the legion from Apulum are known, one of them at Călan. The *signiferi* from the *V*

[304] Berciu, Popa 1964, 307-310.
[305] Băluţă, Berciu 1980, 116-121.
[306] Piso 1993.

Macedonica legion are found in three different locations: at Potaissa, Napoca and Sucidava, but he bulk of them are attested on their list dedicated on the occasion of the upgrade of their *schola* in the legionary fortress. The list is closed by the same number of *tesserarii*; they are equally split between the two legions.

We have placed below these tactical ranks two *conductores armamentarii*, both from Apulum, both of them possibly higher-ranking than simple *immunes*.
We have evidence for an important number of *duplicarii*, 22, 12 of them mentioned on the list of veterans from Potaissa. The placement and nature of this rank are very hard to determine. We might be tempted to attribute it to legionary cavalry, but such position is not found among the *equites legionis*. In this situation we can only say that he benefits from double pay or double food rations, their role being unclear.
As far as legionary cavalry is concerned we have six *equites legionis* obviously enjoying a rank of *principalis* and also increased pay as compared with a humble *miles*. Four of them serve in the *XIII Gemina* while two do so in the *V Macedonica.*

Our list containing petty officers from Dacia comes to an end with a modest group of 24 ***immunes***. Half of them are *custodes armorum*; only three are found at Apulum, the rest being from the legion *V Macedonica.* The only type of military musician found in the legions from Dacia is the *tubicen*; we find five such individuals, almost equally represented in both military centers. We have evidence for three technical posts: two *mensores*, one especially mentioned as *agrimensor* of the legion *XIII Gemina* and an *architectus* of the legion *V Macedonica.* Finally, five simple *immunes* bring this long list to an end; they do not list the reason for their immunity.

Besides these two legions representing the base of the provincial army we know very well the plentiful activity done here by vexillations of the *VII Claudia* legion, especially in the south of the province, even when this territory was not part of the province of Dacia. We find much evidence in the province concerning the petty officers from this legion. One *speculator* and two *beneficiarii consularis*, probably from the *officium* of the governor of Moesia Superior, alongside a beneficiary of one of the legion's tribunes, make up the group of staff officers from this legion. Besides them, the list dedicated to the *VII Claudia* legion is completed by three *principales*, *evocatus, optio* and *signifer* and a *medicus.* Still, we must mention that part of them cannot be attributed to this legion with 100% certainty but mostly because of the place of discovery.

Three other legions are also attested in Dacia through their petty officers. The legion *IIII Flavia*, for a short time included in the army of Dacia, is known through a *frumentarius* eventually promoted to centurion; *VI Victrix* is also known because of a *frumentarius* found at Sarmizegetusa and from *XV Apollinaris*, which took part in the Dacian wars and populated the before-mentioned city, we know one *signifer*. Finally, a beneficiary of the consular is found at Crăiunel, Harghita County, his legion of origin not mentioned.
Last but not least we must mention C. Iulius Fronto, *scrinarius praefectorum praetorii*, who represents Dacia in the *cohors I Praetoria*, in Rome and serves as *immunis*; this represents an important evolution of the province and a level of openness unheard of before (during the Severan dynasty).

In general, the discussion on the origin of these ranking soldiers is simple. They are all citizens, as they are legionnaires, but in several cases, one can trace their recent citizenship grant or their *cognomen* can indicate a not so Roman origin. This state of facts simply confirms the provincialization of recruitment already noticed by a series of works concerning the matter.

I.3. Auxiliary units

I.3.1. Lower hierarchy of auxiliary units

Auxiliary petty officers share a great diversity. The *alae quingenariae* had 16 decurions commanding the *turmae* and the *milliariae* ones 24. The cohorts have 6 or 10 centurions and the mixed ones have between 3 – 6 decurions and 6 – 10 centurions. As far as *numeri* are concerned, the number of petty officers varies. The *navarchi* and *trierarchi* of the fleets are positioned under legionary centurions. In auxiliary units we find pretty much the same posts and ranks as in legions, except that because they are not as well documented, some are missing. Plus, we find different categories of functions specialized on certain troops.[307] The system is comparable to that of the petty officers from the legions, the ones from the *auxilia* being paid one and a half or twice the basic salary, as proven by their own titles – *sesquiplicarius* and *duplicarius*.[308]

In fact, the lower hierarchy of the auxiliary troops was shaped after the legionary one.[309] The posting of officers or petty officers of auxiliary troops from among legionaries or citizens was a custom, at least at the beginning of the Principate, with the goal of giving a solid Roman aspect to all troops serving the Empire. The result is that the feeling of membership in the Roman world that these auxiliaries felt was much evolved on both sides as can be understood from Strabo or Josephus, authors who refer to the auxiliaries as 'Roman cohorts'.[310]

Curiously, when analyzing military diplomas we find evidence of few ranks, besides the *pedites* and the *equites*: *centurio* (2), *decurio* (6), *sesquiplicarius* (2). We find no *optio* or *singularis* or any other *immunis* mentioned on Paternus' list. From this point of view diplomas offered to those in the navy are more generous, attesting: *ex remigibus classicis, ex armorum custode, ex principale, ex optione, ex sesquiplicario* and *ex librario sesquiplicario*.[311]

Alae

Because the *equites singulares Augusti* is a cavalry troop it borrowed its name, organization and obviously hierarchy from the classic auxiliary cavalry troop, the *ala*.[312] We know from ancient sources that the *principales* of the *ala* are *decurio, duplicarius, and sesquiplicarius*.[313] Alongside these three primary ranks others are later mentioned.[314]

The commander of the subunits of the *ala* was the *decurion,* who might be appointed from the lower ranks of the legion, where he could also return, this time as a centurion. In the *Digest*[315] we find mentioned the duties of such a petty-officer: keeping the soldiers in the camp, mobilizing them for instruction and training, taking care of the keys from the gates, inspecting the guards while on duty, inspecting the distribution of food and sick men, physical punishments, receiving complaints.[316] Caesar was the first to recruit specialized cavalry from among the Gauls and the Germans, to organize it in *turmae* and to appoint citizen decurions in command, chosen of course from legionaries.[317] First among the decurions is the *decurio princeps*. His main task is to head the *tabularium alae*.[318] We have many examples of decurions promoted from *eques legionis* as well as from lower ranks in the troop, *sesquiplicarius* or *duplicarius alae.*[319]

Duplicarius is ranked lower than the *beneficiarius consularis*. In this post, alongside that of decurion, were appointed often legionary soldiers or *principales* and *decurio alae* could become besides legionary centurion even commander of an auxiliary unit.[320] The move to a post of *duplicarius alae* represented a promotion for the simple legionary, resulting that it implied a significant increase in the salary.[321] The rare mentions of *sesquiplicarius* indicate the fact that most often he was promoted.[322] The post might simply signify the increase in pay. The tasks of these two ranks cannot be established from their names, the only certain fact being that they benefit from increased payment. Hypothetically, it can be stated that the *duplicarius* was the second in command of a decurion of a *turma*, thus taking over the role of the *optio* from the infantry. It is only in the case of riders and medics that the mention *duplicarius* indicates double payment; when it is adjacent to another rank the term means only that the person in cause receives double food rations.[323]

Previous studies[324] showed there might be four types of standard bearers in the *alae*: *signifer alae*, *imaginifer*,[325] *signifer turmae* and *vexillarius*. Evidence is few and hard to discern, just like the titles.[326] It is certain that the *imaginiferi*, *vexillarii* and *signiferi* are found in all three types of auxiliary troops and that the *ala* had more than

[307] Le Bohec 1994, 46, 57.
[308] Southern 2006, 123.
[309] Southern 2006, 334.
[310] Saddington 1982, 188-189.
[311] Roxan 1995, 145-146.
[312] Domaszewski 1908, 53.
[313] Hyginus, *De munit.castr.* 16.
[314] Dixon, Southern 1992, 25.

[315] 49.16.12.2.
[316] Webster 1969, 147-148.
[317] Speidel 1980, 212-213.
[318] Domaszewski 1908, 53. Speidel 1981 opts for the awarding of this special title (*princeps*) only in the case of a special assignment, with an *ad-hoc* command.
[319] Gilliam 1958, 159-160.
[320] Domaszewski 1908, 54. Apparently, the inscription was misread, cf. Mann 1993.
[321] Cheesman 1914, 35; Speidel 1970.
[322] Domaszewski 1908, 54.
[323] Sander 1959, 247.
[324] Domaszewski 1908, 56.
[325] Same role as in the legion cf. Cheesman 1914, 40; Dixon, Southern 1992, 25.
[326] Domaszewski 1885, 56; Webster 1986, 106.

one type of flag, *signa* probably for the *turmae* and *vexillum* for the whole troop.[327]

Just like any unit commander, the *praefectus alae* (or *tribunus*) has a specialized staff of functionaries, gathered around him and at his disposal. The *officium praefecti alae* is lead, like any other such bureau, by the *cornicularius*, who coordinates personnel.[328] Most likely he is a *principalis*, like all categories of *cornicularii*, even receiving double pay,[329] although according to some opinions it is the other way around, having only a simple *immunis* status.[330] What is truly certain is the fact that if he is a *principalis*, the *cornicularius alae* is the only such rank in the *officium praefecti alae*. The adjacent position might be occupied by the *act(u)arius alae*, both of them being assisted by a number of *librarii*. Probably either one of the two takes care of the *tabularium alae*, whose existence is only supposed.[331] The personnel serving in such an office are completed with a series of posts of *immunes*: *strator*, *stator*, *beneficiarius*,[332] *librarius*, the strange *optio equitum* from Raetia, or *singularis*,[333] along with the *secutor alae* and *eques pecuarius alae*.[334] Other posts of *immunes*, not in an *officium* are: *curator turmae*,[335] *custos armorum*, *librarius*, military musicians, *tubicen*, *bucinator*, *cornicen*[336] and technicians *mensor*[337] and *medicus alae* (obviously with the special status already discussed).[338]

Auxiliary cohorts

Centurio princeps is first among the centurions, probably having the same role as the *decurio princeps* from the *alae*, that of coordinating the troop's *tabularium*.[339] A peculiar situation can be observed at Căşei, where we have two such *principes*, serving in the same cohort.[340]

The position of commander of a *centuria* from an auxiliary cohort, *centurio*, ranks above *beneficiarius consularis*, *imaginifer legionis* and *beneficiarius procuratoris*. The holder promotes directly to legionary centurion.[341] The major difference between legionary centurions and auxiliary centurions rests in the manner of appointment to the position and in the previous status of the candidate. In Egypt, *praefectus Aegypti* appoints auxiliary centurions or decurions while legionary centurions are appointed directly by the emperor.[342] In the case of this type of centurions it is noticeable that, quite often, they are appointed on this post as soon as they are enlisted.[343] Either they are enlisted directly on this post, coming from a family with an exceptional military tradition or with exceptional training or they occupied a short series of posts, promoting rapidly due to ability and favor. This kind of appointment on the post directly from enlistment was attractive for the sons of local elites (even in cities of veterans, like in Dacia), for the sons of professional soldiers, even centurions, who have priority when promoted, or for barbarian nobles from communities where an auxiliary unit is recruited.[344] Like in the case of *decurio alae* or *duplicarius alae* we possess evidence concerning the promotion of legionary soldiers to auxiliary centurions, but not prior to occupying a tactical rank in the legion. In the 3rd century the term *ordinatus* is coined with reference to the centurion of the cohort.[345] *Decurio princeps* leads the chancellery of the riders from the *cohortes equitatae*. *Decurio cohortis* ranks below *decurio alae*, but above *beneficiarius consularis* and *beneficiarius procuratoris*. Legionary soldiers can also be appointed on this post. Usually they promote inside the unit, and peregrine decurions can become legionary centurions only after receiving citizenship following *honesta missio*.[346] A possible hierarchy of these centurions has been considered on the basis of rosters of *cohors XX Palmyrenorum*, thus obtaining a certain supposed seniority, but the situation cannot be extrapolated upon.[347]

Principales of *pedites*: *imaginifer*, tactical ranks *(signifer, optio, tesserarius)*, similar with other troops.[348] *Immunes*: *tubicen*, *cornicen*, *bucinator*, together *aeneatores*, following the model of the legion. Initially they did not have *cornicines*, thus the status of heavy infantry is not confirmed, but the situation changes after they are allocated.[349] The technicians are *mensor* and *medicus*.[350]

Principales of *equites*: *imaginifer* is apparently attested among cavalrymen of the cohorts,[351] *vexillarius*

[327] Alexandrescu 2010, 169, 171.
[328] One of the few evidence for the existence of this post is CIL III 7651, from Gilău. Domaszewski 1908, 55.
[329] Breeze 1971, *passim*.
[330] Cheesman 1914, 41, 43.
[331] Stauner 2004, 173.
[332] Nelis-Clément 2000, 111-113. Sources do not mention the number of such *beneficiarii* from an auxiliary troop but based on the analogy with the legion we might suppose that a *praefectus cohortis/alae quingenariae* had five while the commander of a *milliaria* one had 12. No career for such beneficiary is recorded, only one is probable, from *eques* to *strator* then *beneficiarius tribuni cohortis*. The possible promotions might be awarded inside the same *officium*, to *cornicularius* or in provinces containing legions one might hope for a transfer. Ott 1995, 36-37.
[333] Domaszewski 1908, 55-56, also confirmed by Chessman 1914, 41, 43-44 and Dixon, Southern 1992, 25.
[334] Domaszewski-Dobson 1967, xvii.
[335] Apparently specific for cavalry, just like *summus curator*. Webster 1969, 148; Speidel 1992c, 138.
[336] Only the fleets, cohorts and *alae* have military musicians of all three categories, probably among the *immunes*. The place and function in the army of the *lituus* and *liticines* are not clearly known. Webster 1969, 148; Alexandrescu 2010, 67, 75-76.
[337] As compared with the only *mensor cohortis* known by Domaszewski (CIL XIII 6538), other inscriptions attesting these technicians in auxiliary troops have not been found, with the exception of a *mensor* and a *discens mensorem* on *papyri* from Dura. Their numbers and role in the auxiliary troops is not known. Sherk 1974, 551.
[338] Chessman 1914, 41, 43-44 and Dixon, Southern 1992, 25.
[339] Domaszewski 1908, 56.
[340] CIL III 7631.

[341] Domaszewski 1908, 56.
[342] Gilliam 1958, 158.
[343] Gilliam 1958, 164: *BGU* 696, I, 20-25, CIL VIII 9389 (*decurio alae*) and 21034 (*centurio singularium*).
[344] Gilliam 1958, 165-166.
[345] The same CIL III 7631 from Căşei.
[346] Domaszewski 1908, 57.
[347] Fink 1947, 167, 170; Fink 1953, 210-213; Davies 1976, 253-254, 267.
[348] Webster 1969, 149.
[349] Domaszewski 1908, 58; *contra* Cheesman 1914.
[350] Domaszewski 1908, 58.
[351] CIL III 3256.Domaszewski 1908, 58, *contra* Fink 1947, 169.

confirming that they are organized in the same manner as those from the *alae*. As far as military musicians are concerned, they have only the *bucinator*.[352] The status of the simple *eques* is open for debate, in the same manner that the status of the *eques legionis* is compared to that of *pedes*. The only clue as to them having a special status, besides the difference in payment, is the presence of *schola equitum*.[353] In *cohortes equitatae* men were not directly enlisted as riders but they reached this status after serving between 2-5 years as infantrymen.[354] Later on, posts of a technical role, *immunes*, similar to those from the legions or from the troops in Rome are found, inscriptions [355] indicate a *capsarius* and *papyri* from Dura add *sacerdos*, *aedituus*, *magister campi* or *cibariator*.[356]

Officium praefecti/tribuni cohortis. It is likely that the posts in the staff of auxiliary cohorts were identical to those of the legionaries, but on a reduced scale:[357] *cornicularius* again seems to be the only post of *principalis* from this staff, assisted by *actarius* who ranks above *adiutor cornicularius consularis*, *beneficiarius*, *and librarius*.[358] A more unusual post is that of *magister cohortis*, perhaps an instructor or replacement of an *optio*.[359] Also, an inscription[360]refers to *stratores tribuni cohortis XX Palmyrenorum*.[361]

Numeri

The centurions in the *numeri* usually have Roman names or the possession of citizenship is indicated despite their non-Roman origins. In some cases they might originate in the population from where the unit is recruited; in this case we cannot find centurions of British origin. As far as their titles are concerned attempts have been made to distinguish between *ordinatus* and *ordinarius* but there seems to be no difference, otherwise the forms would not be abbreviated to *ordin.*, and the extra word *princeps* would not be added. The *optiones* seem to assist centurions but they are to be found only in units originating in the East which are found in Dacia or Africa and probably are more Romanized as far as organization and hierarchy are concerned, than their counterparts in Germany. It was believed that the petty officers from the *numeri* came from the legion, in order to help and train these illiterate barbarians, but evidence for this is uncertain. Also, by analyzing the *grafitti* found in their camps and the names of those petty officers we can deduce that at least some of them were literate. When associated with other auxiliary units petty officers might be transferred from those units; such is the case for the *vexillarii Africae et Mauretaniae Caes. qui sunt cum Mauris gentilibus*.[362] Evidence concerning the promotion of simple *milites*, like the examples for the other troops, are few in numbers.[363] *Pedites*. *Centurio* ranks below the cohort's centurion because one of the latter is appointed *praepositus numeri*. In the 3rd century they might be called *ordinarius* or *ordinatus*.

Signifer,[364] *optio*. Domaszewski believes that the post of *tesserarius* as well as those of military musicians are missing, according to previously mentioned reasons.[365] *Equites*. *Decurio*, *duplicarius*, *sesquiplicarius*, *vexillarius*, *imaginifer*, *baiulus* (a kind of messenger).

There is of course an *officium praepositi*, similar to others of the type. From the category of *immunes* - technicians we find the *librarius numeri*, *mensor frumenti numeris*, or *magister numeri* who is in fact *eques alae*, thus an instructor.[366]

Equites and *pedites singulares* have the same organization and ranks as the *ala* does, respectively *cohors*. To be precise, the following are known: in the case of *equites* – *decurio princeps*, *decurio*, *duplicarius*, *sesquiplicarius*, *summus curator* or *curator*, *actarius*, *optio*, *praeco*, *stator*, *stationarius*. In the case of *pedites* – *centurio*, *optio*, *signifer*; or in general – *vexillarius*, *aedituus*, *librarius*, *beneficiarius*.[367]

The Roman military fleet follows the organizational and hierarchical principles of the land army. We encounter a few *principales* typical for the fleet: *pausarius*, *pitulus*, *velarius*, *proreta*, *gubernator*, *nauphylax*. Others are identical to those from the land army: *optio*, *suboptio*, *armorum custos*. We find technical *immunes*: *subunctor*, *medicus*, *naupegus*, *caementarius*, *dolabrarius*, *faber*, *tubicen*, *bucinator*; but also petty officers from the staff: *cornicularius*, *adiutor*, *scriba*, *tesserarius*, *signifer*. The pay structure is the same: *miles*, *immunis*, *sesquiplicarius*, *duplicarius*.[368] As correspondents of centurions from the land army in the fleet we encounter *navarchus* and *trierarchus*. Antoninus Pius equates them to centurions from the legions. *Trierarchus* seems to outrank *optio peregrinorum* and *decurio*, because they qualify directly for the post of centurion, after such a character[369] was previously *decurio*, then becomes *trierarchus* and *primuspilus*.[370]

The career of the auxiliary soldier

In the period before the Flavian dynasty only 13 auxiliary centurions were attested who had citizenship as compared

[352] Domaszewski 1908, 58.
[353] Domaszewski 1908, 58.
[354] Davies 1971; Roxan 1995, 141.
[355] ILS 9169.
[356] Domaszewski-Dobson 1967, xvii.
[357] Webster 1969, 149.
[358] Stauner 2004, 173.
[359] Domaszewski 1908, 58-59.
[360] AE 1931, 116.
[361] Domaszewski-Dobson 1967, xvii.
[362] CIL XVI 108.
[363] Southern 1989, 100-102.
[364] There also are auxiliary *signa*, composed only of zoomorphic representations. They are not Roman, probably belonging to barbarian *numeri*. Domaszewski 1885, 74-75.
[365] Domaszewski 1908, 60. Contrary to Domaszewski about illiteracy or inferiority Domaszewski-Dobson 1967, xvii.
[366] Domaszewski 1908, 60-61.
[367] Speidel 1978, 31-32.
[368] Reddé 1995, 151. Attested personnel in provincial fleets from Britannia and Germania: *trierachi, navarchi, centuriones, gubernatores, proretae, nautae* and *classiarii*. Saddington 1990, 230-232.
[369] CIL X 3342a.
[370] Reddé 1995, 152.

with seven that were peregrines. Later however, when Roman citizenship will spread further, the number of peregrine centurions will lower considerably.[371] A good example for this is the one from Pannonia. Out of the 16 auxiliary centurions who are known all of them have Roman citizenship. There probably existed also peregrine centurions but in this case they are not proven. One can reach this post in two ways, both including some posts of *principales*, in auxiliary troops and legions. The leading centurion of a cohort is *princeps*, one of the *principes* from Pannonia being promoted to *exercitator equitum legionis*, then serving two terms as centurion in Rome. The situation for the decurions of the *ala* is similar, just as is the case for their hierarchy. 40 such *decuriones alae* are attested, in the 1st century part of them being peregrine and then in the 2nd-3rd centuries most of them being citizens. This rank was obtained after passing through several *principales* posts. The 14 decurions of the *equitatae* cohorts are all citizens and share the same careers with their centurion colleagues. Also, most *optiones, vexillarii, cornicularii* and *duplicarii* are citizens, and if almost all *signiferi* are citizens, both *imaginiferi* are peregrines. *Tesserarii* appear only in cohorts, their role in the *alae* being taken over by the *sesquiplicarii*, most of them being peregrines. Still, it can be noticed that their sons already enlist in the legions.[372]

Perhaps one of the most interesting and important military careers in the Roman army is that of Ti. Claudius Maximus. The relative recently discovered inscription from Philippi, Macedonia, sheds a new and relevant light over the career of a cavalryman. It is so important because it develops both inside a legion and in the ranks of the auxiliary cavalry. Ti. Claudius Maximus was probably born in Philippi, Macedonia and his first post is that of *eques legionis VII Claudiae.* This fact is at least a little strange because a legionary was enlisted as *miles*, then, if he turned toward the cavalry forces, he became *discens equitum* (apprentice cavalryman) then *factus eques*. If he was enlisted directly as a cavalryman he must have fulfilled superior criteria or come from a family with a strong military tradition. Careers that might indicate previous posts of *miles* and *discens equitum* are found in inscriptions.[373]

The next post in his career is expressed by the words *factus quaestor equitum*, attestation in premiere. If these legionary riders are part of the regular *centuriae* it is only normal that they are paid from these funds. This post confirms however the fact that they had a special fund, destined perhaps only for the maintenance of horses and equipment.[374] The next post held by Maximus started a wave of controversies: *singularis legati legionis eiusdem*, again a post attested for the first time. It was believed[375] that the guards of the legionary legate are *equites legionis*. It is obvious that these *singulares legati legionis*, to the point that they exist in other places, rank higher, perhaps having more important duties.[376] After Speidel, Rankov and Pavkovic got involved in the discussion following the discovery of this inscription.

From this point on he becomes *vexillarius equitum*, obviously a post of great reputation. From here on the career becomes interesting, the character in question being transferred in the auxiliary cavalry: *factus duplicarius a divo Traiano in ala secunda Pannoniorum.* This promotion continues the Republican tradition of appointing auxiliary petty officers from among the legionaries in order to ensure competence and rigor of the auxiliary troops. He is appointed by the emperor himself, though normally this should have been done by the governor; this means that at this point Trajan is in command of the army, perhaps even during the Dacian wars. It is obvious that his pay was also increased and we can state that *duplicarius alae* had a higher salary than a *sesquiplicarius legionis.*[377] His career finally ends with *et ab eodem factus decurio in ala eadem*. Promotion from *duplicarius* to *decurio alae* is very often encountered. The simple *eques legionis* could also become decurion of an *ala*, afterwards moving on to centurion in a legion. In spite of many decorations and acts of bravery the career of Ti. Claudius Maximus is very slow and complicated.[378]

There is no evidence for the existence of a rigorous promotion system in the auxiliary troops, especially because we have in the whole Empire only four careers with three posts and one with four. The *papyri* from Dura contribute to our level of understanding but do not reveal the complete mechanism because they are too close in time and the multitude of auxiliary troops prevents us to decipher an organized system.

Most careers are grouped around the posts of *singularis consularis*, *sesquiplicarius*, *duplicarius* and *decurio*. From *singularis consularis* one could promote to *vexillarius*, *duplicarius alae*, or directly to *decurio* (three cases). This position inside the governor's staff might offer them an advantage for promotion to centurion. From this post it was moved on to *in officium* and then to *vexillarius*. One can promote to *vexilarius* directly from *eques* or *imaginifer*. From *eques* one promotes to *sesquiplicarius*, directly to *duplicarius* or *decurio* and *duplicarius* promotes directly to *decurio.*[379]

Two careers are more special: Iulius Dexter[380] – *eques, curator turmae, armorum custos, signifier turmae*;

[371] Roxan 1995, 143.
[372] Lőrincz 2001, 122-128, 141.
[373] ILS 9090, CIL VI 3409, CIL XI 5646 = ILS 2081, CIL XII 2602 = ILS 2118. Speidel 1970, 143. Once this inscription was found legionary cavalry is attested under Domitian, making obsolete the theory stating that it did not exist between Vespasian and Hadrian. At the latest, in AD 66 the legion *VII* Claudia is transferred from Dalmatia to Viminacium, in Moesia.
[374] Speidel 1970, 144.

[375] Starting with Domaszewski, who attributes the legionary *singulares* to the tribune.
[376] Speidel 1970, 144.
[377] Speidel 1970, 145-146. This promotion raises an interesting discussion regarding auxiliary pay compared to that of the legionaries.
[378] Speidel 1970, 149.
[379] Breeze 1974, 282.
[380] CIL VIII 2094=ILS 2518.

Maximus[381] – *pedes*, *eques*, *singularis cohortis*, *singularis alae*, *tesserarius*, *librarius*, *actarius*, *cornicularius*. It seems that these careers prepared the men in question for becoming legionary centurions because the rest stop at the post of decurion and the owners cannot afford very imposing monuments on which to present their careers. It can be stated that the most important trait of auxiliary careers is flexibility.[382]
The two careers presented previously indicate the possibility for a complicated latter, somewhat logical that ended up with the post of legionary centurion. Most end with the post of *decurio* that one can reach from a multitude of posts, different as status. Both *immunes* – *singularis consularis*, as well as *principales* – *sesquiplicarius*, *stator*, however only four soldiers receiving double payment, promote directly to decurion. The situation is different from the one in the legions where only *eques*, who does not receive double pay, becomes directly a centurion. This fact indicates the advantages of *singulares* and *stratores* who represented the auxiliary elite of a provincial army and were clearly favored.[383]

The *papyri* from Dura might help deduce the length of service for petty officers although the problems are relative and must be discussed separately for infantry and cavalry. For cavalrymen the *stipendia principalis* is between 12-20 years, with the lower margin for the *sesquiplicarius* and those serving over 20 years had probably lost their chance to promote to a post of command. *Pedites* could promote, besides to *eques*, to *dromedarius*, who also are enlisted in the registers of the *centuria*. In their cases selection for promotion was done much quicker than in the case of cavalrymen and those advancing to centurions did so between their 14th and 20th year of service, just as in the case of legions. It is difficult to exactly state for how long a post of *principalis* was held, probably for at least three years, because in the rosters of the cohort *XX Palmyrenorum* most hold the same posts in the three-year span.[384]

I.3.2. The lower hierarchy of auxiliary units from Dacia

Alae

Principales. The *principales* post best represented in the *alae* from Dacia is that of *decurio*. He leads a squadron of auxiliary soldiers, *turma*, and is the highest rank of auxiliary *principalis*. He cannot be equated with centurions because he promotes to legionary centurion, this being the ending point of a soldier's career. We know 29 decurions from the *alae* of Dacia.

The best-represented troops as far as attestations of decurions are concerned are *ala I Tungrorum Frontoniana* and *ala II Pannoniorum*. From the unit in Ilişua we have attested no fewer than seven decurions. Probably created since the time of Augustus, from the Germanic tribe of the Tungrians, it was included in the army of Lower Germany, until the upheaval of the Batavi, when it is brought in Dalmatia. Then, during the time of Hadrian it is brought from Pannonia in Dacia, initially in Banat, afterwards at Ilişua.

No. **1** is the deceased to whom the funerary monument is dedicated. He was a veteran, former decurion, lived for 50 years and is buried alongside his infant son, Valerius Vitalianus. The monument is set up by his two other sons, both of them *principales* in the same unit. This soldier was discharged after 25 years of honorable service, from the post of decurion. The military tradition would have been inherited in the family, as noticeable from the careers of his other sons. No. **2** is one of the sons of no. **1**, who sets up the monument dedicated to his father and his brother. He is decurion in the same unit his father served in. We notice that the military tradition is passed on; in the present case great skill in mounted combat is also involved, the son starting off an advanced position thanks to his father's career.

No. **5** sets up a temple for himself and his family, *salute genio sancto scolae decurionum*. The inscription mentions *schola decurionum,* the headquarters of an association of petty officers. As he is the founder of the temple he might even be *decurio princeps*, the most important decurion in an *ala*, and commander of the *tabularium alae*.[385] *Turma* of no. **6** is attested on a *tabula ansata.* Based on the period when the troop arrived in Dacia it was established when it was in Banat.[386] Finally, the reconsidered no. **7** was promoted decurion of the unit after serving in the *equites singulares* of the governor.[387]

Ala II Pannoniorum is attested in Syria, in AD 88, and then arrives in Moesia Superior. It takes part in the Dacian and Parthian wars then returns in Dacia Porolissensis, at Gherla. Perhaps the most reputed soldier of this unit, decurion and *duplicarius*, is Ti. Claudius

[381] AE 1926, 74.
[382] Breeze 1974, 282.
[383] Breeze 1974, 283.
[384] Breeze 1974, 284-286.

[385] The position of *decurio princeps* as chief of the *tabularium alae* is uncertain.
[386] Petolescu 2002, 78.
[387] See Cupcea 2013a.

Maximus, from Philippi,[388] the captor of Deceballus. The unit arrives in Moesia Superior and takes part in the Dacian wars, shortly afterwards remaining in the new province. Five decurions are attested from this unit.

No. **11** and **12** are two decurions found on a funerary monument. The first is the deceased and the second is the dedicator, probably former colleagues. Again a *collegium* of these petty officers is indirectly mentioned, as well as its role in the burials of its members.

No. **9** is a very important character. A veteran, former decurion of the unit from Gherla he finds himself in a very privileged position. He is a decurion in Sarmizegetusa, in the *municipium* Napoca and in the *kanabae* of the *XIII Gemina* legion at Apulum. It is from this position that, along with his family and with his own money, he sets up a votive plaque for the Great Mother of the Gods, for the health of the emperor, in a place provided by decree of the decurions.

No. **8**, a veteran, former decurion in the same *ala II Pannoniorum* dedicates an honorary marble slab for the emperors Severus and Caracalla and for the *Caesar* Geta, whose name is erased (dated AD 198-211).

Ala I Batavorum ∞, from Războieni-Cetate has five attested decurions, two of whom are uncertain: two unknown soldiers, one at Potaissa and the other at Ocna-Mureş-Salinae and two with their names known, one at Potaissa, the other one at Războieni. Regarding the two men from Potaissa we cannot say for certain that they belonged in the *ala I Batavorum* ∞ but we assume this fact because of other inscriptions connected with the troop found Turda and because of the proximity of the two points. Some of the greatest specialists in mounted warfare, the Batavians, known for their special abilities, such as swimming while mounted (according to Tacitus), have formed elite cavalry troops, being preferred even by Trajan when he instituted his guard of *equites singulares Augusti*.[389] This unit, one of the few *alae milliariae* from the Empire, passed through Germania and Pannonia Superior, being part of the garrison of Dacia Superior for a period of time.

Among them, no. **15**, veteran, former decurion, probably from the unit from Războieni is the father-in-law of the veteran Valerius Maximus, centurion in the legion *V Macedonica* and dedicates with his parents-in-law and his children the funerary monument of his wife, Valeria Maximilla. No. **17**, Illyrian by origin and heir to a deceased comrade, was recently reconsidered as decurion of this unit, and not of the *equites singulares*, but promoted and transferred from among them.[390]

Ala Siliana, with its four-decurion representatives,[391] takes third place in this ranking. The troop stationed at Gilău, decorated many times, *bis torquata, bis armillata, civium Romanorum*, has a long history. Created by C. Silius, proconsul of Africa in the AD 60, called in in Italy during the Civil War from AD 69, it took part in putting out the rebellion of the Batavians, at this moment being also decorated. It was later transferred in Pannonia, it also receives the distinction *civium Romanorum*; it takes part in the Dacian wars and is garrisoned in the new province, later on taking part in Trajan's campaign against the Parthians. No. **16** dedicates along with his decurion colleagues a votive altar in honor of Iupiter *optimus maximus*. No. **21** had his funeral monument recently discovered in Napoca. He is the head of a military family, from which also nos. **47** and **20** (from the auxiliary cohorts) are part of.[392]

Three auxiliary units were stationed at Micia. The *ala I Hispanorum Campagonum* ∞ has two decurions attested. The Hispanic troop is not present in any of the diplomas before AD 114 but it is possible it was recruited before AD 69. It was sent in the Balkans, in Dalmatia, then in Pannonia Inferior, the best-documented period in its existence being when it was stationed at Micia.[393] The decurions attested from this troop are nos. **22** and **23**.

Ala I Claudia nova miscellanea, initially made up of Gauls and Hispanics, is garissoned in Dalmatia, takes part in the conquest of Thrace then is taken to Germania to combat the Batavian uprising Later moved to Moesia, it is mentioned among the troops in Dacia in the diploma of July 2nd, AD 110 from Porolissum (CIL XVI 163=IDR I DiplD 3), but it does not station here. Still, we have two decurions from its ranks, both of them veterans, attested in Dacia, no. **24** and **25**.

Ala I Bosporanorum serves in Syria during the 1st century then, with the occasion of the Dacian wars, is brought in Moesia from where it will move on in Dacia after the conquest, being camped in Dacia Superior. We have a single decurion mentioned, no. **26**, from the time the unit was stationed at Micia.

Ala I Hispanorum is recruited during the 1st century, probably by Augustus, camped in Germania, moved under the Flavian dynasty in Pannonia, then Moesia, finally camped in Dacia Inferior, at Slăveni. No. **27**, attested on the lower fragment of a votive altar, might just as well be a decurion from Romula.

Ala numeri Illyricorum is known under several names: *vexillatio equitum Illyricorum*, *numerus equitum Illyricorum*. Initially it is a vexillation from the *numerus*

[388] Speidel 1970, *passim*.
[389] Speidel 1994, 38.
[390] See Cupcea 2013.

[391] Isac 1997, 19-20. He includes among the decurions T. Aelius Veteranus, improbable. In fact the whole inscription is problematic, apparently two deceased persons being mentioned but in fact only the second one has the age of death written down and the fact that his son dedicated the monument. Possibly a single character. Also, the author (Isac 1997) does not mention Aurelius Umbrianus.
[392] Piso, Tecar 2013.
[393] Petolescu 2002, 72-73.

of *equites singulares* coming from Illyricum, camped at Brâncoveneşti, in Dacia Superior and organized as an *ala*. Decurion no. **28** is attested.

The only *principalis* to be part of the staff of an *ala* commander ever attested in Dacia is a *cornicularius*. From this post he leads the *officium tribuni alae Silianae*, from Gilău. No. **30** is one of the few such officers attested in the whole of the Empire.[394]

The most prestigious post of *principalis*, outside of any superior officer's staff is the *imaginifer*. The situation is the same for auxiliary cavalry; from this post it was possible to promote directly to decurion or centurion. Most of the times however the reputation and long term of the post prevented the soldier from being further promoted.

We have four *imaginiferi alae* attested in the whole of Dacia, two of them from the same unit, *ala I Hispanorum Campagonum*, from Micia, one from an unknown unit, from the north of the province and one from *ala II Pannoniorum*.

No. **31** is an *imaginifer* in the ala from Micia, where he served for 20 years. This number confirms the hypothesis mentioned above, that the post was held towards the end of the career and that further promotion was rare. No. **34** coming from around Samum reveals briefly his status as *[imag]in(i)f(er) al[ae?]*, without further mentions. The soldier might originate in one of the troops nearby, either *ala I Tungrorum Frontoniana*, or *ala II Pannoniorum*.

Going down the hierarchy ladder we reach the rank of *duplicarius*. The decurion's second in command of a cavalry troop received double pay compared to the simple *eques*, as noticed in the name of the post, and promoted, most often, to decurion. A number of four *duplicarii alae* are attested in Dacia and the same is true for the *sesquiplicarii*; all of them come from five different units.

No. **35** sets up a marble bas-relief depicting a knight and a priest performing a sacrifice for the association of *duplicarii* from the *ala I Hispanorum*, with the attribute *optimum collegium*. The mention of such an association is interesting, confirming the fact that any category of *principales* might organize themselves of an *ad-hoc* base in such organizations, with charitable goals of course.

No. **36**, *duplicarius* from *ala I Tungrorum Frontoniana*, is the son of the decurion from no. **1**, the first one to be mentioned in our list, for whom he sets up, along with his brother, Valerius Laetillus, decurion in the same unit, the funerary monument. As stated before, they complete a family with a strong military tradition. Another interesting fact is that he is a subordinate of his borther: *duplicarius decuriae eiusdem*, which basically excludes any family restrictions inside the same unit.

The same number of four *sesquiplicarii alae* is known in Dacia. Two of them are part of the same unit camped at Slăveni, the other two coming from the north of the province. No. **40** entitles himself *sescupluc[arius]*, a form of *sesquiplicarius*, on a marble votive relief fragment that probably depicted Minerva.

Of the flag bearers from auxiliary cavalry units we find only four *signiferi* and a *vexillarius*. All come from the army of Dacia Porolissensis, two of them coming from the same unit.

No. **46** was a *signifer* in *ala Siliana*; he dies however at the age of only 29 years, 11 months. His parents, Aurelius Reburus, a veteran of the same troop and Fabia dedicate his funerary monument. Recently the only *vexillarius alae* of Dacia has been discovered in Napoca, that is no. **47**, the son of no. **21**, decurion of the same unit, bearing a Greek cognomen.[395]

No. **45** is *signifer alae (II Pannoniorum) ac cu[rator]*. His name indicates his Thracian origin and he probably inherits the anonymous rider for whom he sets up the monument. It is probably one of the few examples of auxiliary careers from Dacia. Before becoming *signifer* we can assume that Dines was a *curator*, an *immunis*, technical post, charged with stable maintenance. The post is typical for cavalry troops and such a career would have meaning.

Finally we discuss a lost inscription, probably from Slăveni, that survived as a copy on a sheet of paper. It describes the dedication of an altar to Sol, for the health of the soldiers of the *ala I Hispanorum*; it is set up by the *principales*. It is a new example of association between the petty officers of this troop for charitable purposes, as mentioned before. This closes the list of *principales* from the *alae* of Dacia; the remaining posts attested in the province can be included among the *immunes*.

For the cavalryman of an *ala* one of his ideals that he might easily fulfill during his career was to become *eques singularis*, in the cavalry guard of the governor. We have three such riders who, because they were detached to the consular *praetorium* from Apulum, implicitly received a privileged status in their unit of origin, respectively that of *immunis*. However we cannot say for sure if they were initially detached from the *ala* mentioned, or simply transferred as *equites* after the completion of their service in the guard.[396]

Nos. **48-50** are three such examples, of horsemen that were previously *singulares*. If they were elected in the guard from the same unit, we cannot know, the only sure thing being that they resumed regular service as *alae* cavalrymen, which is not necessarily a promotion. The next guard post attested, that of *stator*, is part of the unit commander's staff, in the *immunes* category, no. **51.**

[394] Only one more seems to be attested, AE 1926, 74, in Isauria. Cf. Stauner 2004, 173.

[395] Piso, Tecar 2013.

[396] Detailed discussion in Cupcea 2013a.

A series of scribes and secretaries are needed in the *officium* in order to keep evidence of the troop's documents. The most numerous of these are *librarii*, the basic form of secretary. We have two *librarii alae* attested in the *alae* in Dacia, one at Slăveni, the other at Ilişua. They are joined by an *immunis*, possibly another *librarius*,[397] from Gherla.

[397] Watson 1965, *passim*.

	Name	Rank	*Ala*	Source	Place
1	C. Val. Vitalis	*decurio*	*I Tungrorum Frontoniana*	CIL III 811	Ilişua
2	Valerius Laetillus	*decurio*	*I Tungrorum Frontoniana*	CIL III 811	Ilişua
3	Aur. De(?cimus)	*decurio*	*I Tungrorum Frontoniana*	CIL III 802	Ilişua
4	Val. Valerianus	*decurio*	*I Tungrorum Frontoniana*	CIL III 805, 808	Ilişua
5	P. Ael. Paulinus	*decurio*	*I Tungrorum Frontoniana*	CIL III 7626	Ilişua
6	Val. Firmus	*decurio*	*I Tungrorum Frontoniana*	CEpR III 221	Pojejena
7	P. Ael. Atilianus	*decurio*	*I Tungrorum Frontoniana*	AE 1913, 54=ILD 795	Negrileşti
8	C. Ant. Crispinus	*decurio*	*II Pannoniorum*	CIL III 1375=IDR III/3 54	Micia
9	T. Fl. Longinus	*decurio*	*II Pannoniorum*	CIL III 1100=IDR III/5 253	Apulum
10	Sept. Pisusus	*decurio*	*II Pannoniorum*	AE 1975, 732=ILD 588	Gherla
11	Secundinus	*decurio*	*II Pannoniorum*	CIL III 12542	Gherla
12	M. A[ur?]	*decurio*	*II Pannoniorum*	CIL III 12542	Gherla
13	*Unknown*	*decurio*	*I Batavorum ∞*	CIL III 933a=IDR III/4 68	Salinae
14	Atilius Celsianus	*decurio*	*I Batavorum ∞*	CIL III 933=IDR III/4 72	Războieni
15	P. Aelius Tertius	*decurio*	*I Batavorum ∞ ?*	ILD 511	Potaissa
16	*Unknown*	*decurio*	*I Batavorum ∞ ?*	CIL III 12546	Potaissa
17	Bersius Ingenuus	*decurio*	*I Batavorum ∞*	IDR III/5 522, 312	Apulum
18	Sex. Val. Saturninus	*decurio*	*Siliana*	CIL III 845	Gilău
19	T. Aelius Paulus	*decurio*	*Siliana*	CIL III 846=7650	Gilău
20	Aur. Umbrianus	*decurio*	*Siliana*	CIL III 864=7663	Gilău
21	Aur. Maximus	*decurio*	*Siliana*	Piso, Tecar 2013	Napoca
22	Quintianus	*decurio*	*I Hispanorum Campagonum*	AE 1903, 68=IDR III/3 133	Micia
23	C. Val. Antestius	*decurio*	*I Hispanorum Campagonum*	CIL III 1383=IDR III/3 186	Micia
24	C. Iul. Verecundus	*decurio*	*Claudia nova miscellanea*	CIL III 14216[7]=IDR II 43	Drobeta
25	Ulp. Metius	*decurio*	*Claudia nova miscellanea*	CIL III 7871=IDR III/3 184	Micia
26	Fir(?) Florentinus	*decurio*	*I Bosporanorum*	CIL III 7888=IDR III/3 246	Germisara
27	Domitius Rufus	*decurio*	*I Hispanorum ?*	AE 2002, 1224=ILD 131	Slăveni
28	Ulpius Dassius	*decurio*	*numeri Illyricorum*	AE 1992, 1473=ILD 444	Brâncoveneşti
29	T. Aelius	*decurio*	*?*	CIL III 1179=IDR III/5 472	Apulum
		Officium praefecti/tribuni			
30	Aurelius Carinus	*cornicularius*	*ala Siliana*	CIL III 7651	Gilău
		Principales			
31	Censorinus	*imaginifer*	*I Hispanorum Campagonum*	CIL III 7870=IDR III/3 172	Micia
32	Quintus	*imaginifer*	*I Hispanorum Campagonum*	IDR III/3 183	Micia
33	Brisenus	*imaginifer*	*II Pannoniorum*	Lupa 11349	Gherla
34	*Unknown*	*imaginifer*	*II Pannoniorum /Tungrorum ?*	ILD 784	Dej
35	Iulius Marinus	*dupl.*	*I Hispanorum*	AE 1963, 125=IDR II 505	Slăveni
36	Val. Saturninus	*dupl.*	*I Tungrorum Frontoniana*	CIL III 811	Ilişua
37	Verecundus	*dupl.*	*Siliana*	CIL III 848	Gilău
38	Aurelius ?	*dupl.*	*milliaria ?*	CIL III 7644	Sutor
39	Aur. Aurelianus	*sesq.*	*I Hispanorum*	IDR II 506	Slăveni
40	*Unknown*	*sesq.*	*I Hispanorum*	IDR II 507	Slăveni
41	Aur. Valens?	*sesq.*	*numeri Illyricorum*	AE 1992, 1472=ILD 440	Brâncoveneşti
42	Aur. Vitellianus	*sesq.*	*I Tungrorum Frontoniana*	CIL III 791	Ilişua
43	*Unknown*	*signifer*	*I Tungrorum Frontoniana*	CIL III 6274=IDR III/1 107	Vršac (SRB)
44	*Unknown*	*signifer*	*I Tungrorum Frontoniana*	CIL III 813	Ilişua

45	Dines	*signifer curator*	*II Pannoniorum*	AE 1993, 1329=ILD 594	Gherla
46	Aur. Fabius	*signifer*	*Siliana*	CIL III 847	Gilău
47	Aur. Musaeus	*vexillarius*	*Siliana*	Piso, Tecar 2013	Napoca
	Collective	*principales*	*I Hispanorum*	IDR II 510	Slăveni
		Immunes			
48	*Unknown*	*sing. cos.*	*numeri Illyricorum*	AE 1988, 947=IDR III/5 631	Apulum
49	Dasas	*sing. cos.*	*I Batavorum*∞	CIL III 7800=IDR III/5 522	Apulum
50	Mucasenus	*sing. cos.*	?	IDR III/5 558	Apulum
51	Aur. Tato	*stator*	*II Pannoniorum*	ILD 590	Gherla
52	*Ignotus*	*librarius*	*I Hispanorum*	CIL III 14216[34]=IDR II 532	Slăveni
53	Aur. Themaes	*librarius*	*I Tungrorum Frontoniana*	CIL III 804	Ilişua
54	*Ignotus*	*immunis*	*II Pannoniorum*	AE 1971, 397=ILD 593	Gherla

Table I.5. *Immunes* and *principales* of the *alae* in Dacia.

Auxiliary cohorts

The highest-ranking post of ***principalis*** from the auxiliary cohorts is that of centurion. This post is different from the one in the legions because of several characteristics. First of all because the status of its holder. An auxiliary centurion could be a man with a modest social status, while his legionary counterpart was part of local elite. Secondly, in order to become legionary centurion an exceptional military career, and eventually *evocation*, were needed. Last but not least auxiliary centurions promote to legionary centurions or in a cohort in Rome.

In Dacia we find 20 auxiliary centurions. Two of them are *centurioni principes*, the most important ones in a cohort, probably coordinating the activity of the *tabularium cohortis.*

The unit with the highest number of attested centurions is *cohors I Flavia Hispanorum* ∞. This unit is found at the end of the 1st century in Moesia Superior, it takes part in the Dacian wars and remains in the occupation army of the province. Here it is mentioned for the first time on the milestone from Aiton, then is mentioned at Buciumi and Orheiul Bistriței.

Four bronze tablets, fashioned as *tabulae ansatae*, with the inscription punched in, attest a number of four *centuriae* and their centurions, from *cohors I Flavia Hispanorum* ∞ - nos. **4-7**.

The second auxiliary unit in Dacia in number of centurions is *cohors II Flavia Commagenorum*, from Micia. The troop is mentioned at the end of the 1st century in Moesia Superior, it takes part in the Dacian wars and remains in the occupation army of the province. It is mentioned for the first time during the reign of Hadrian and will occupy the garrison from Micia until the disappearance of the province. Three centurions are attested. No. **11** is burried at Sucidava, probably his place of origin; he dies at 33 years of age, after serving for 15 years. The character enjoys a good career, at the age of only 33 becoming centurion in this unit, with plenty of future opportunities.

Cohors I Vindelicorum ∞ *equitata* is initially mentioned in Pannonia and then, in AD 100, in Moesia Superior; it takes part in the Dacian wars and remains in Dacia. Initially camped at Vărădia (Arcidava), it is moved at Tibiscum. We have three centurions attested from this unit.

No. **10** is attested on an oval bronze plaque, considered an identification document by the authors of IDR, of the soldier Iulius Martialis, from the cohort *I Vindelicorum*, belonging to his *centuria*. It is dated in the time frame when the troop stationed at Vărădia, before moving to Tibiscum. On a marble slab, containing a list of soldiers and their ranks, from Tibiscum, an unknown centurion (no. **8**) is attested, along with five colleagues, decurions, attributed to this unit because of the names.

No. **9** is a very important character in the Dacian colony. A veteran, former centurion in the cohort *I Vindelicorum*, former *IIvir* of the colony Sarmizegetusa, he reached the respectable age of 89; three of his children are present on his monument. We are dealing with an honorable family with origins in Palmyra one of the many active ones from Dacia who are part of local elites, especially in Sarmizegetusa.

Next we find a number of seven centurions from just as many auxiliary cohorts from Dacia. *Cohors III Campestris* is a troop with a special history. Before the Dacian wars it is attested in Moesia Superior, then it moves in Dacia at the beginning of the province and afterwards is mentioned at the middle of the 2nd century in Moesia. The diploma from 1st of April 179, from Drobeta, again mentions it among the troops from Dacia. Thus, we only know for sure that the troop was initially charged with guarding the two bridgeheads from Drobeta, the problem of its movement being in fact an issue pertaining to the authority of Moesia's governor.[398]

At Pojejena, also in Banat, stationed *cohors V Gallorum*, a troop that passes from Moesia following the conquest of Dacia and remains here. On a votive marble slab fragment the sign of the centurion (>) was deciphered, but the soldier remains unknown, no. **15**.

The cohort *I Flavia Commagenorum* was part of the army of Moesia Inferior, even during the time when this province encompassed Oltenia. Following that it remained in Dacia Inferior, where it had a good building activity, along legionary vexillations, as proven by the discovered tile stamps. At some point in the 3rd century archaeological discoveries indicate that it stationed in the fortification from Câmpulung-Jidova.[399]

Cohors I Ubiorum was part of the army of Moesia Inferior, being attested many times at Capidava. In Dacia it is mentioned in diplomas and on tile stamps from Odorheiu Secuiesc, Ozd or Târnăveni.[400]

In the north of the province three units attest the same number of centurions. *Cohors I Brittonum* ∞ *Ulpia torquata p f c R* earns its titles and decorations after effecting bravery acts during the Dacian wars. Camped at Porolissum, it is also attested at Bologa or Buciumi.

Cohors I Batavorum ∞ comes along with the legion *II Adiutrix* in 86 from Britannia in Pannonia. Following the conquest of Dacia it will station in the fortification from Romita-Certiae. No. **18**, a centurion of this cohort, dies at the age of 27, after 9 years of service. The funerary monument is set up by Florius Virilis, veteran, former decurion in the same unit; the deceased is his nephew. It is a special career of a precocious man who becomes centurion very young.

398 For the whole discussion see Piso, Benea 1987, *passim*.
399 Petolescu 2002, 96.
400 Petolescu 2002, 125.

Finally, *cohors I Britannica ∞ c R equitata* moves from Pannonia to Moesia Superior at the end of the 1st century. After participating in the Dacian wars it remains in the army of Dacia Porolissensis, camped in the fortification from Căşei-Samum. No. **3**, who originates from Virunum, Noricum, veteran, former centurion in this troop, dies at the age of 60 and is buried at Apulum, his wife Aelia Aestiva taking care of his monument. No. **20**, recently revealed on his father's funerary monument, no. **21** (of the *alae*) is a centurion in an unknown unit, but considering the place of discovery, he could be part of the cohort of Samum.[401]

The cases of the two petty officers entitling themselves *princip(?)* were left behind. The continuation *princip(ibus)* is possible, meaning that the two are *principes* centurions of this troop,[402] or *princip(ales)*; in the second case we are dealing with a generic name for petty officers.[403] It is certain that the two set up an altar for the *genius* of a *schola ordinatorum*, the headquarters of an association, apparently composed of the centurions and decurions of the cohort.[404] Thus, Domaszewski's proposal seems to be more plausible, although he admits that the doubling of such a rank is peculiar.[405]

It is a well-known fact that inside *cohortes equitatae*, cavalry contingents, also called *turmae*, are commanded by decurions. The relation between them and centurions from the same unit is identical with the one between legionary centurions and the decurions of the *alae*. Thus, the promotion of a decurion from an *equitata* cohort was to the post of centurion in the same unit. In Dacia we find a number of 11 decurions of cohorts, no fewer than six belonging to the same unit, *cohors I Vindelicorum*.

No. **21**, veteran, a former decurion of this unit, is attested on the epitaph written in hexameters on his monument, on a white marble slab; he presents here his long military career. The monument is very important for the study of the incursion of the Latin language in superior artistic forms in Dacia.[406]

On the marble slab fragment IDR III/1 129, also from Tibiscum, that contains a list of soldiers and their ranks, including the above-mentioned unknown centurion, five colleagues, decurions, are also present; they were attributed to this unit based on their names, nos. **22-26**.

We know a single decurion from the cohort *II Flavia Commagenorum*, from Micia, no. **29**. *Cohors I Batavorum ∞* presents, in the monument of the centurion C. Campanius Vitalis also one of his relatives, decurion, no. **28**, an uncle or grandparent of the centurion C. Campanius Vitalis; this relative sets up the funerary monument.

Cohors I Hispanorum veterana is mentioned in Moesia Inferior, at the beginning of the 2nd century and it takes part in the Dacian wars according to *P. Hunt.* In Dacia Inferior it is stationed at some point also at Arutela. No. **27** is mentioned on a round silver plate, with the inscription inside a *corona*.

Cohors II Hispanorum is attested in the 1st century in Illyricum, afterwards in Pannonia. It takes part in the conquest of Dacia and in Trajan's Eastern campaign. After returning from the Orient it remains stationed in Dacia Porolissensis, at Bologa.

Cohors IIII Hispanorum equitata appears only in the diplomas from Dacia Superior, from the middle of the 2nd century onwards, being stationed at Inlăceni.
We encounter only two of the tactical ranks that activate in a *centuria*, the *optio* and *signifer*. *Optiones* are to be found in only two cases in the cohorts from Dacia.

The unit from Dacia with the highest number of auxiliary centurions, *cohors I Flavia Hispanorum ∞*, also includes an *optio*. No. **33** dedicates a bronze votive hand to Iupiter *optimus maximus* Dolichenus, discovered in Myszkow, Ukraine, but most likely coming from Orheiul Bistriţei and taken as spoils of war following an invasion of the *Costobocae*.[407]

This ends the list of the *principales* serving in the auxiliary cohorts from Dacia. The following nine sodiers occupy posts from the *immunes* category.

Most of them serve in the superior commanding officer's staff, whether prefect or tribune, the most prestigious among them being the *beneficiarii*.

No. **36**, an Illyrian according to his name, is *beneficiarius tribuni cohortis III Campestris*, dies at the age of 40 and we know that his funerary monument is set up and cared for by his inheritor, Linda Severus, also Illyrian.

We know a number of four *actarii* who serve in the cohorts from Dacia. They coordinate the writing of the *acta* of the unit, because they are members of the *officium praefecti/tribuni*. It is typical of these *actarii* to dedicate altars to Minerva, probably the patron goddess of their associations. No. **39**, presumably from *cohors I Alpinorum*, that seems to have stationed for a while in Dacia, takes care of the funerary monument of one of his colleagues, a legionary in *V Macedonica*, *exceptor* of the legate.

In the marble fragment pertaining to no. **42** we find that there are two posts written down: *actarius*, probably for one of the cohorts in the area, either from Micia or Tibiscum and *magister k(?ampi)*, a riding instructor. If this restoration is correct, his promotion is special, from a

[401] Piso, Tecar 2013.
[402] Domaszewski 1908, 56.
[403] Isac 2001, 45-46.
[404] Domaszewski 1908, 57.Especially because in the 3rd century auxiliary centurions are called *ordinati.*
[405] Domaszewski 1908, 56.
[406] Petolescu 2002, 141, includes him in *numerus Palmyrenorum Tibiscensium*.
[407] Petolescu 2002, 112.

position of *immunis*-technician to one of superior secretary, the main redactor of the documents of a troop.

Librarii are also included in the secretary category. We have to *librarii* attested but we must note that one of them calls himself *immunis*; they are nos. **40** and **41**. The first of them also notes down a promotion, from *eques cohortis I Hispanorum*, to *librarius*.

Finally, *custos armorum*, a very important *immunis* post, also for future promotions, is occupied by two soldiers serving in the cohorts from Dacia.

No. **43** is a veteran, former *custos armorum* of the cohort *I Vindelicorum* and decurion of the colony Sarmizegetusa, Finally, no. **45** is *miles cohortis I Sagittariorum in figlinis magister super milites*, mentioned in the raw paste of a brick. It is quite clear that the character is nothing more than an *immunis*, a technician, charged with coordinating work at the brick workshop.[408]

[408]There is an opinion stating that in the 3rd century it gradually replaces the term *optio* but in the present case we could not make it equivalent to this rank, or at least not the variant of *optio centuriae*. Breeze 1976, 129. *Magistri* are mentioned as holding military posts similar to *optiones*: *magister ballistariorum*, *magister fabr(um, icae?)*, *magister calcariorum*, *magister figlinarum*, *magister kampi*, *magister equitum*, *magister barcariorum*.

	Name	Rank	Cohort	Source	Place
		Principales			
1	Caecilius Aelianus	*> princeps*	*I Britannica ∞ c R eq*	CIL III 7631	Samum
2	Ti. Aur. Iulianus	*> princeps*	*I Britannica ∞ c R eq*	CIL III 7632	Samum
3	P. Ael. Tertius	*centurio*	*I Britannica ∞ c R eq*	AE 1980, 751=IDR III/5 484	Apulum
4	Valerius	*centurio*	*I Flavia Hispanorum ∞*	AE 1983, 851a=ILD 640	Buciumi
5	Severinus	*centurio*	*I Flavia Hispanorum ∞*	AE 1983, 851b=ILD 644	Buciumi
6	C. Pomp. Valerianus	*centurio*	*I Flavia Hispanorum ∞*	AE 1983, 851c=ILD 645	Buciumi
7	L. Turanius Arte(mid?)orus	*centurio*	*I Flavia Hispanorum ∞*	AE 1983, 851d=ILD 646	Buciumi
8	*Unknown*	*centurio*	*I Vindelicorum*	IDR III/1 129	Tibiscum
9	P. Ael. Theimes	*centurio*	*I Vindelicorum*	CIL III 1472=IDR III/2 369	Sarmizegetusa
10	Clemens	*centurio*	*I Vindelicorum*	IDR III/1 110	Vărădia
11	Ceselinius	*centurio*	*II Flavia Commagenorum*	AE 1959, 32=IDR II 205	Sucidava
12	Crispus	*centurio*	*II Flavia Commagenorum*	CIL III 7850=IDR III/3 88	Micia
13	C. Iul. Martialis	*centurio*	*II Flavia Commagenorum*	CIL III 13733=IDR III/3 177	Micia
14	L. Hera(clius?)	*centurio*	*I Flavia Commagenorum*	IDR II 365	Romula
15	*Unknown*	*centurio*	*I Galorum. et Pannoniorum*	IDR III/1 15	Pojejena
16	Aur. Equester	*centurio*	*I Ubiorum*	AE 1980, 862=ILD 437	Odorheiu Secuiesc
17	Ael. Firmus	*centurio*	*I Brittonum*	ILD 697	Porolissum
18	C. Campanius Vitalis	*centurio*	*I Batavorum ∞*	CIL III 839	Romita
19	P. Ael. Papirianus	*centurio*	*III Campestris*	CIL III 14216^{10}=IDR II 44	Drobeta
20	Aur. Manes	*centurio*	*?*	Piso, Tecar 2013	Napoca
21	P. Ael. Ulpius	*decurio*	*I Vindelicorum*	CIL III 8001=IDR III/1 157	Tibiscum
22	Proculus	*decurio*	*I Vindelicorum*	IDR III/1 129	Tibiscum
23 to 26	4 x *Unknown*	*decurio*	*I Vindelicorum*	IDR III/1 129	Tibiscum
27	Terentius	*decurio*	*I Hispanorum veterana*	CIL III 12603=IDR II 581	Arutela
28	Florius Virilis	*decurio*	*I Batavorum ∞*	CIL III 839	Romita
29	L. Sossius	*decurio*	*II Flavia Commagenorum*	CIL III 1355=IDR III/3 105	Micia
30	Aelius Tato	*decurio*	*II Hispanorum*	CIL III 843=627	Bologa
31	Aelius Fuscus	*decurio*	*IIII Hispanorum*	AE 1988, 974=IDR III/4 283	Mihăileni
32	C. Speratus	*optio*	*I Britannica ∞ c R eq*	CIL III 824	Samum
33	Caius	*optio*	*I Hispanorum ∞*	AE 1998, 1113=ILD 805	Orh. Bistr.
34	Aelius Certus	*signifer*	*I Batavorum ∞*	CIL III 13760	Potaissa
35	Aur. Proculinus	*signifer*	*III Delmatarum*	AE 1973, 463=IDR III/1 87	Tibiscum
		Immunes			
36	Liccaius Vinentis	*beneficiarius tribuni*	*III Campestris*	IDR II 45	Drobeta
37	Aur. Candidianus	*actarius*	*I Vindelicorum*	ILD 198	Tibiscum
38	Ianuarius	*actarius*	*II Flavia Commagenorum*	IDR III/3 111	Micia
39	M. Val. Maximinus	*actarius*	*I Alpinorum*	ILD 561	Napoca
40	Val. Valerianus	*eques/librarius*	*I Hispanorum ∞*	IDR II 582	Arutela
41	Catullus	*immunis*	*II Flavia Commagenorum*	IDR III/3 123	Micia
42	M. Aur. Valens	*actarius/magister k?*	*?*	IDR III/2 270	Sarmizegetusa
43	Aur. Laecanius Paulinus	*custos armorum*	*I Vindelicorum*	IDR III/1 137, 138	Tibiscum
44	Aelius Verus	*custos armorum*	*II Flavia Commagenorum*	IDR III/3 120	Micia
45	Aurelius Mercurius	*magister*	*I Sagittariorum*	IDR II 107	Drobeta

Table I.6. *Immunes* and *principales* of the cohorts in Dacia.

Numeri

In theory, the *numeri* follow the hierarchic structure of the Roman army, respectively of auxiliary troops, with but few exceptions. These exceptions are more related to their status of *nationes* than to the generic name of *numeri.* The lowest category of centurion, it is still held by soldiers who have citizenship, be they honorably discharged and *evocati* or transferred from other units, even from *principales* posts.

In the *numeri* from Dacia we encounter a number of eight centurions, four of them belonging to the same unit, *numerus Palmyrenorum Porolissensium* and one probably a *princeps.*

No. **4** is probably a decurion of Napoca, of Syrian descent. No. **5** is *augur* of the *municipium Septimium Porolissensis.*

No. **7** was the head of a family of Syrian-Palmyrean descent, with a rich military tradition. A veteran, former centurion in *numerus Palmyrenorum O?*, he lived for 40 years. Given his nickname, Maximus, we might conclude that he was an exceptional soldier, probably becoming centurion after serving in the ranks of that unit. His funerary monument is set up by his family: the son, Aelius Septimius Romanus, a soldier in the *XIII Gemina* legion, also a petty officer, *adiutor offici cornicularioram*, the daughter, Septimia Septimina *qui et* Revocata, the wife, Cornelia Antonia, his freedman Septimius Asclepiades, *augustal* of the colony, having taken care of the consecration.

No. **1** is mentioned on an altar dedicated to the goddess Placida Regina, by the cavalrymen of the *numerus Surorum*, encamped here for a while, through two representative petty officers, this Proculus *(centurio) princeps* and a Caius, *optio.* Contrary to what the authors of the IDR claim the positions of *princeps*, or even that of the *optio*, are not 'insignificant'.[409] I tend towards centurion instead of decurion because of the assistance of this *optio.*

As far as cavalry or mixed units are concerned the ranks of decurion is positioned either first or second after the centurion. In the *numeri* from Dacia we have a number of eight attested decurions, one of them *princeps* and one, maybe three belonging to the *equites singulares.*

No. **9**, *decurio princeps* of *numerus Palmyrenorum Tibiscensium*[410] is attested on a brick found in the pavement of a building inside the fortification, perhaps the *tabularium numeri*, whose leader he is, where it is mentioned the fact that he took care of the building of a portico for the god (perhaps from a temple but unlikely inside a fort).

The *equites singulares* accompanying the consular at Apulum are made up from a selection of *immunes*, chosen from the auxiliary troops inside the province. These *equites* already have a privileged status among their comrades but they are not petty officers; after they finish their mission at Apulum they return to the troops of origin. Many times they come back and serve in higher posts, of *principales*, this distinction smoothing out their path towards more substantial promotions. Because it is a unit created *ad-hoc* at the governor's headquarters, this type of *numerus* has its own petty officers. It is obvious that the *turmae* are led by decurions. We know one or three such officers, unfortunately very badly and attested on fragmentary monuments, nos. **10-12**.

No. **13**, *decurio* or centurion of one of the *numeri* garrisoned in Tibiscum, has been a cause of debate between scholars, being considered a *beneficiarius*, former *singularis.* After several recosiderations, he is believed to have been a decurion promoted from the *equites singulares consularis.*[411]

No. **16** appears on a collective funerary monument belonging to a family with military tradition from Porolissum. He is the head of the family, a veteran, former decurion, who lived for 85 years. The epitaph mentions also Aurelius Passer, no. **25**, *signifer*, dead at 33 years of age, Aelia Sura, dead at 18 years of age and another Aurelius, a soldier in the same unit. Aurelia Sabina and Aurelius, *actarius* – no. **31**, set up the monument. All mentioned soldiers are part of the *numerus Palmyrenorum Porolissensium*, have Syrian descent and three of them serve as petty officers in the unit. Also, the presence of a decurion proves the mixed character of the troop.

The following ranks, in descending order, are the tactical ones, in the case of infantry, or those of *duplicarius*, *sesquiplicarius* or *vexillarius*, in the case of cavalrymen.

Two *duplicarii* are attested in the *numeri* from Dacia.

No. **18** is a *duplicarius* in the *numerus equitum singularium* that sets up three altars, two of them identical in terms of text, both for Mars and Victoria. He presents a short career, for a long time not noticed, due to the abbreviation *S C.* Considered in several instances as *singularis consularis*, it has been reconsidered as *summus curator*, probably the head of all the *curatores*, petty officers with duties regarding stable maintenance in cavalry units.[412]

No. **17**, a veteran, former *duplicarius*, probably in one of the units that camped in the area at a certain moment, perhaps *numerus Surorum,* is found on a collective funerary monument, where he is listed first; he lived for

[409] Florescu, Petolescu, in IDR II, 152.

[410] Piso, Benea 1999, 106. Schmidt-Heidenreich 2000, 299, discusses the possibility that he might be a centurion, by comparison with Proculus (CIL III 1590a=8029; IDR II 338, Romula), assisted by an *optio*.

[411] Piso 1994, 207-209 and Speidel 1994b, 215-217.

[412] Speidel 1978, 53-56, especially 56; Speidel 1992, 137-139; Davies 1976, 137-144.

60 years and next to him rest Aelius Titianus, a veteran, probably his brother, who lived for 70 years and Aelius Valerianus, a son, who lived for 50 years; the monument is set up by the sons Aelius Candidinus, *tesserarius* of the legion *XIII Gemina* and Aelius Veteranus. The military tradition is so deeply rooted in this family that the offsprings have *cognomina* derived from military terms – Veteranus.

We encounter two more ranks that are found exclusively among the cavalrymen from *numeri*: a *vexillarius* and a *sesquiplicarius*. Nr. **28**, *vexillarius* in the *equites singulares*, promoted inside the guard, lived for 32 years, the monument being set up by his inheritor Valerius, *librarius*, probably serving in the same unit.

Nr. **29**, *sesquiplicarius* of *numerus equitum Illyricorum*, stationed on the Upper Olt river,[413] receives citizenship and *conubium* according to the diploma dated July 19th, AD 146. His name might be Thracian but that of his father, Turus, seems Illyrian.
As far as the careers of infantrymen serving in the *numeri* are concerned we have a number of eight soldiers holding ranks that can be included in the *principales* category; all of them are tactical ranks, five *optiones* and three *signiferi.*

In *numerus Palmyrenorum Tibiscensium* we encounter three of the *optiones*. From among them, no. **23**, *optio* in *numerus Surorum*, appears on the altar dedicated to the goddess Placida Regina, along with no. **1**, *centurio princeps*. No. **25**, *signifer* of *numerus Palmyrenorum Porolissensium*, dead at 33 years of age, appears on the collective funerary monument that belongs to the family with a strong military tradition led by the decurion no. **16**.

No. **27**, *signifer* and *quaestor* in *numerus singularium Britannicianorum*, sets up a votive altar for the holy nymphs, patrons of the thermal waters from Germisara, who helped him evade death in AD 186. The association of these two posts is strange. It is known that the *signifer* also carries fiscal duties, in the case of legions - he holds along with the flags the economies of the soldiers serving in his *centuria*. But, in the case of auxiliary troops this extra duty is unknown and unaccepted. Because we are dealing with a troop that evolved from the *equites singulares* it might be assumed that the need for these *signiferi* also charged with financial duties was felt; if this is so it was felt that it was needed to explicitly mention this attribute. On the other hand the clearly named position of *quaestor* is only known in legionary cavalry.

Moving on to the *immunes*, we find a number of ten such positions, most of them charged with administrative duties in the *officia* of the troop's commanding officer, a few being *singulares* in the governor's guard.

One of the most prestigious posts of *immunis* that is not included in the *officia* is that of *custos armorum*. We have a single case, no. **37**. The clerks in charge of writing the *acta* of auxiliary units, the *actarii*, are found in three cases, for the *numeri* in Dacia; they belong to three different troops. No. **30** was for a long time thought of the *equites singulares*, but a recent discovery in Germany changes his status. He is therefore an *actarius peditum singularium*, a post attested only in this instance all throughout the Empire.[414] No. **31** is *actarius* in *numerus Palmyrenorum Porolissensium*, and is mentioned on the collective monument belonging to the family with a strong military tradition led by the decurion no. **17**; he is the one taking care of the monument.

Nr. **32**, *actarius praepositi numeri Surorum*, set up along with the *librarii* serving under him, a votive altar for Mithras Sol Invictus. The monument is important for several reasons. First of all it attests an *officium praepositi*, in the case when he commands a *numerus*, secondly it mentions an *actarius* in a rather important position inside this *officium*. Last but not least it attests a number of *librarii*, serving under this *actarius*. Eve if he has an Eastern origin, he is a Roman citizen.

No. **33** is the deceased inherited by the decurion Bersius Ingenuus, of the *ala I Batavorum* ∞. He is an *eques* of the same unit, transferred from the *equites singulares*. They are both of Illyrian origin, and he is a peregrine.

We encounter two more of the *librarii* above-mentioned, besides the collective from Romula. No. **38**, *librarius* of the *equites singulares* takes care of the funerary monument of his comrade, no. **28**, the *vexillarius*; he is his inheritor. No. **40** is entitled *immunis* from the *numerus Surorum sagittariorum*, of Celtic origin.

No. **34** is a particular character, however unknown, but from his fragmentary inscription one can decipher F EX SC, which may or may not be reconstructed as *...[b(ene)]f(iciarius) ex s(ingulari) c(onsularis)...* Peculiar as it may seem, we cannot exclude the possibility that a former *singularis* was transferred and promoted in the *numerus Maurorum Miciensium*, as *beneficiarius* of the tribune.

No. **35** is a former *singularis* becoming cavalryman of the *ala numeri Illyricorum*, and no. **36** is somewhat similar, only of Thracian origin, in an unknown unit.

[413] Petolescu 2002, 132.

[414] Wiegels 2013, forthcoming. Even if the statuette found in Germany has a different place of discovery, the coincidence with the one of Apulum is too obvious. See also Cupcea 2015, forthcoming.

	Name	Rank	*Unit*	Source	Place
1	Proculus	*centurio*	*Surorum*	CIL III 8029=IDR II 338	Romula
2	Martinus	*centurio*	*Surorum ?*	CIL III 14216=IDR II 537	Slăveni
3	Aur. Iustinus	*centurio*	*Palmyrenorum Porolissensium*	AE 1974, 549=ILD 706	Porolissum
4	Ael. Thema[rsa?]	*centurio*	*Palmyrenorum Porolissensium*	AE 1979, 495=ILD 707	Porolissum
5	M. Cocceius Alexander	*centurio*	*Palmyrenorum Porolissensium*	AE 1979, 499=ILD 712	Porolissum
6	Ael. Maenemachus	*centurio*	*Palmyrenorum Porolissensium*	ILD 714	Porolissum
7	P. Ael. Sept. Audeo	*centurio*	*Palmyrenorum O?*	CIL III 1471=IDR III/2 366	Sarmizegetusa
8	M. Aur. Calpurnianus	*centurio*	*singularium Britannicianorum*	IDR III/3 219	Germisara
9	Marius Aurelianus	*dec. princ.*	*Palmyrenorum Tibiscensium*	AE 1999, 1302=ILD 213	Tibiscum
10	*Unknown*	*decurio*	*equitum singularium*	IDR III/5 626	Apulum
11	*Unknown*	*decurio*	*equitum singularium*	CIL III 7787=IDR III/5 405	Apulum
12	Congonius	*decurio*	*equitum singularium ?*	CIL III 1203=IDR III/5 606	Apulum
13	Iul. Gracilis	*decurio / >*	*? Tibiscensium,*	IDR III/5 219	Apulum
14	Aur. Severus	*decurio*	*Maurorum S?*	AE 1975, 723=IDR III/4 259	Rareş
15	T. Valerius	*decurio*	*Maurorum Optatianensium*	CIL III 7695	Potaissa
16	Aurelius ?	*decurio*	*Palmyrenorum Porolissensium*	AE 1971, 389=ILD 704	Porolissum
17	Ael. Valerianus	*duplicarius*	*Surorum*	CIL III 8034=IDR II 352	Romula
18	P. Ael. Crescens	*duplicarius* *singularis*	*equitum singularium*	CIL III 1098$^{1-2}$, 1160 = IDR III/5 248, 249, 375	Apulum
19	P. Ael. Servius	*optio*	*Palmyrenorum Tibiscensium*	IDR III/1 136, 142, 149	Tibiscum
20	Ael. Gurras	*optio*	*Palmyrenorum Tibiscensium*	CIL III 7999=IDR III/1 154	Tibiscum
21	*Unknown*	*optio*	*Palmyrenorum Tibiscensium*	AE 1967, 394=IDR III/1 162	Tibiscum
22	Mucianus	*optio*	*Palmyrenorum Porolissensium*	AE 1960, 219=ILD 690	Porolissum
23	Caius	*optio*	*Surorum*	CIL III 8029=IDR II 338	Romula
24	*Unknown*	*optio*	*?*	IDR II 654	Oltenia ?
25	Aur. Passer	*signifer*	*Palmyrenorum Porolissensium*	AE 1971, 389=ILD 704	Porolissum
26	Aur. Bassus	*signifer*	*Maurorum Optatianensium*	AE 1932, 81=ILD 757	Optatiana
27	P. Ael. Marcellinus	*signifer* *quaestor*	*singularium Britannicianorum*	CIL III 1396=IDR III/3 243	Germisara
28	Ael. Genialis	*vexillarius ?*	*equitum singularium*	CIL III 7799=IDR III/5 477	Apulum
29	Coca	*sesquiplicarius*	*equitum Illyricorum*	AE 2001, 2155=RMD 269	Dacia ?
		Singulares			
30	M. Aur. Sila	*actarius*	*peditum singularium*	AE 1962, 208=IDR III/5 358	Apulum
31	Aurelius	*actarius*	*Palmyrenorum Porolissensium*	AE 1971, 389=ILD 704	Porolissum
32	Ant. Zoilus	*actarius*	*Surorum*	AE 1914, 120=IDR II 341	Romula
33	Dasas	*singularis*	*ala I Batavorum*	CIL III 7800=IDR III/5 522	Apulum
34	*Unknown*	*beneficiarius ?* *singularis*	*Maurorum Miciensium*	IDR III/3 61	Micia
35	*Unknown*	*singularis*	*ala numeri Illyricorum*	AE 1988, 947=IDR III/5 631	Apulum
36	Mucasenus	*singularis*	*?*	CIL III 1195=IDR III/5 558	Apulum
37	Ael. Zabdibol	*custos armorum*	*Palmyrenorum Tibiscensium*	IDR III/1 134	Tibiscum
38	Valerius	*librarius*	*equitum singularium*	CIL III 7799=IDR III/5 477	Apulum
39	*Unknown*	*librarius*	*Maurorum Tibiscensium*	IDR III/1 172	Tibiscum
	Collective	*librarii*	*Surorum*	AE 1914, 120=IDR II 341	Romula
40	Claudius Montanus	*immunis*	*Surorum*	CIL III 8032=IDR II 350	Romula

Table. I.7. *Immunes* and *principales* of the *numeri* in Dacia.

I.3.3. Particularities of auxiliary *immunes* and *principales* from Dacia

Alae

We know of more than 50 characters that hold a rank or activate on a post of *immunis* or *principalis* in the *alae* from Dacia. Among them, the majority are decurions, basic ranked soldiers of auxiliary cavalry; they role is comparable in importance to that of centurions from the legions. They are the connection between the commander and the soldier and they effectively command the movements of the squadrons. With a total number of 29 they are spread in several units from the province, some of them offering a greater number of mentionings; in other cases we find only one of them.

The best-documented hierarchy is encountered in the troop *ala I Tungrorum Frontoniana*. We find six decurions, most of them at Ilişua, only one at Pojejena, probably from when the troop stationed in Banat, before coming to Dacia Porolissensis. Of the seven mentioned four of them are serving and two are veterans. Next we encounter four more *principales*, one *duplicarius*, one *sesquiplicarius* and two *signiferi*. Finally, we encounter only one *immunis*, a *librarius*, serving in the troop.

The second troop as far as the number of attested petty officers is concerned is *ala II Pannoniorum*. Ten characters occupy 11 posts. Most of them are decurions – five to be more exact, three of them attested at Gherla and one at Apulum and Micia each; the former two are both veterans, very important characters in civilian life. Of the ones found at Gherla one is a veteran. Otherwise, the troop from Gherla contains three *principales*, two *imaginiferi* (one is uncertain) and a *signifer*, who previously held the *immunis* post of *curator*. The other two *immunes* are *stator*, respectively *immunis*.

Ala I Batavorum ∞, from Războieni-Cetate, contains a number of seven petty officers. Of them, five are decurions, found at Ocna Mureş-Salinae, Războieni-Cetate, Apulum and two at Potaissa (uncertain). Only one of them is a veteran. A fifth *principalis* who is mentioned is a *duplicarius*, at Sutor-Optatiana. This series ends with a horseman from the governor's guard, *singularis consularis*, found at Apulum.

Ala I Hispanorum, from Slăveni is in the same situation as the previous one. It has five individuals and a collective of petty officers attested; four individuals and the collective are *principales*, all at Slăveni. The *principales* are one decurion, one *duplicarius*, two *sesquiplicarii* and a collective that entitles itself *principales*. A single *immunis*, *librarius*, is mentioned.
Eight petty officers are to be encountered also in the *ala Siliana* from Gilău, all of them *principales*: four decurions – two of them veterans – assisted by one of the very few *cornicularii alae* from the Empire, the single *vexillarius alae* of Dacia, one *duplicarius* and one *signifer*.

At Micia the cavalry troop *ala I Hispanorum Campagonum* has attested four very high-ranking *principales*: two *decurioni* and two *imaginiferi*, all of them found at Micia.

From the *ala numeri Illyricorum* we encounter one *decurion* and one *sesquiplicarius* at its garrison from Brâncoveneşti and one *singularis consularis*, obviously serving away in Apulum.

*Ala Claudia nova miscellanea*has two decurions that are attested; both are veterans, settled and buried in Dacia although the troop only passes briefly through the north-Danubian province.
We encounter one decurion from *ala I Bosporanorum* who goes to Germisara for treatment while the troop stations at Micia.

Cohorts

A total number of 45 petty officers are to be encountered in the cohorts from Dacia. When taking into account the duality of this type of troops it is obvious that both infantry and cavalry troops are attested, each with its own petty officers. The overwhelming majority of these soldiers - 31 people - are either centurions or decurions. It is quite understandable that these two ranks are the most important ones in the lower hierarchic structure of auxiliary cohorts and that most military careers aim towards them. Also they bridge the gap towards the legions, by becoming centurions there.

The best-documented unit is *cohors I Vindelicorum*, from Tibiscum; this might be so because it is a mixed one, *equitata*. Three centurions are mentioned, each in a different location: Tibiscum, Sarmizegetusa and Vărădia. As far as the cavalry is concerned six decurions are mentioned, all of them at Tibiscum. One *actarius* and one *custos armorum*, both of them from the *immunes*, belong to this troop.
Cohors II Flavia Commagenorum from Micia, with seven known petty officers takes second place in our list. The only *principales* among these seven soldiers are three centurions, one at Sucidava and two at Micia and one decurion, also at Micia. The other three are *immunes*, *actarius*, *custos armorum* and *immunis*, all at Micia.

The unit from Căşei, *cohors I Britannica ∞ civium Romanorum equitata* has a number of five known petty officers, four of them being centurions. Two of them, *principes*, attested on the same monument, at Căşei, present a situation that is at least strange as observed even by Domaszewski: the doubling of the post of *centurio princeps* in a troop. The third, attested at Apulum, has a remarkable civilian career. Besides them another *optio* is encountered, also at Căşei.
Another unit, *cohors I Flavia Hispanorum* ∞, from Buciumi, has four known centurions; all of them are attested on small *tabulae ansatae* found where the troop was stationed.

In Dacia Porolissensis, at Romita, *cohors I Batavorum* ∞, has three attested petty officers: one centurion and one decurion, both at Romita – this also proves the dual nature of the unit – and one *signifer* at Potaissa.
Cohors I Hispanorum veterana ∞ has one decurion and one *librarius*, both attested at Arutela. Also, *cohors III Campestris* leaves evidence of its activities at Drobeta through two petty officers: one centurion and one beneficiary of the tribune, perhaps the only one of its kind to be found in Dacia.

Three other centurions are attested in the cohorts *I Gallorum et Pannoniorum*, at Pojejena, in *I Ubiorum*, at Odorheiu Secuiesc and in *I Brittonum*, at Porolissum. Also there are two decurions in *II Hispanorum*, at Bologa and *IIII Hispanorum*, at Mihăileni. From the ranks of the *principales* we encounter one *optio* from the cohort *I Hispanorum* ∞, at Orheiul Bistriţei and one *signifer* from *III Delmatarum*, at Tibiscum.

Three *immunes* remain to be discussed: one *actarius*, presumably from the cohort *I Alpinorum*, one *magister*, head of the brick workshops belonging to the cohort *I Sagittariorum* at Drobeta and one *actarius* who also seems to have held the post of *magister k(?ampi)*, at Sarmizegetusa; he probably comes from one of the cohorts nearby.

Numeri

As far as the *numeri* are concerned, we encounter a number of 40 petty-officers, 28 of them being posts of *principales*, most of them centurions and decurions but the percentage is not as high as in other types of auxiliary troops.

The best-documented unit from of the point of view of military hierarchy is the *numerus Palmyrenorum Porolissensium*, with eight characters, only one of them being an *immunis*, *actarius*. *Principales* are: four centurions, one decurion, one *optio* and one *signifier*, all encountered at Porolissum.

The next unit in this list is the *numerus equitum singularium*, the governor's mounted guard, from Apulum. No fewer than three decurions are mentioned. One *duplicarius* and one *vexillarius* complete the list of *principales*. One *immunis*, a *librarius* also belongs in the guard.

Next comes *numerus Surorum* with two centurions, one of them a *princeps*, one *duplicarius* and one *optio*, three at Romula and one at Slăveni. The unit also includes a number of *immunes*: one *actarius* with all the *librarii* serving under him and one *immunis*, all at Romula.

Numerus Palmyrenorum Tibiscensium is present in our list because of its five petty officers: one *decurio princeps*, three *optiones* and one *custos armorum*, all at Tibiscum.

Numerus Maurorum Optatianensium has two *principales* who are attested, one *decurio*, at Potaissa and one *signifer* at Sutor. So does the other unit of Moors, the *numerus Maurorum Tibiscensium*, with a *singularis* at Apulum and one *librarius* at Tibiscum. *Numerus singularium Britannicianorum*, also contains two *principales*: one centurion and one *signifier* who is also a *quaestor*; both of them are at Germisara.

Four units have one such character attested. *Numerus Palmyrenorum O?* has one centurion at Sarmizegetusa, *numerus Maurorum S?* has one decurion at Rareş, *numerus equitum Illyricorum* has one *sesquiplicarius* and *numerus Maurorum Miciensium* one *beneficiarius*, former *singularis*. Also, there is an unknown *optio*, discovered in Oltenia who might be from some *numerus* from the area.

Regarding the origin of the lower auxiliary officers from Dacia, a few things can be stated. First of all, for the cavalry, the vast majority are citizens, evidently, recently made, but even so, the only clear examples of peregrines are standard bearers (rather odd), an *imaginifer* and a *signifer*, and two *singulares*. For the citizens, most of them date from after AD 212, and an important part preserves their peregrine *cognomina* as *signum originis*. In the auxiliary cohorts the situation is similar, only that no actual peregrines can be detected between *immunes* and *principales*. Finally, the *numeri* make no difference, only a *sesquiplicarius* bearing a simple peregrine name. All these statistics lead us to the conclusion that even in auxiliary units, one vital condition for promotion was citizenship. The time of its grant can only be previous of the enlistment, therefore the perspective of promotion being rather reduced for the peregrines.

I.4. The role of *immunes* and *principales* from Dacia in the provincial society

What is the mark that the petty officers left on the society of the province? Besides their contribution to the good administration of the province and the keeping of public order in the cities and on the roads, a category of petty officers participates in the social life of the province.[415] We have a very small number of such characters attested as veterans and those who seem to play a tangible part in municipal life are the *beneficiarii*. Of this group only the *beneficiarii consularis* can truly enjoy special status because they activated in the governor's bureaucratic apparatus. Just four such beneficiaries are attested as occupying municipal functions: two of them are *decuriones* and *IIviri* in Sarmizegetusa, one is a decurion at Apulum and finally one, while in service, as *aedilis* at Napoca.[416] Besides them only two other soldiers are exceptionally honored by being accepted in the *ordo decurionum*, as *decuriones supranumerarii*, while still serving. They are two of the sons of the former beneficiary C. Iulius Valerius, decurion and *IIvir* at Sarmizegetusa; one of them is a *frumentarius* in *XIII Gemina*, probably sent in the Empire with some highly sensitive task and the other is praetorian, even *scrinarius* of the prefect of the *praetorium*.[417]

We have proof that military medical personnel could include valuable men, with superior training and status. This is evidenced by the doctor of the legion *VII Claudia* being honored with *ornamenta decurionalia* at Drobeta. Besides these all other petty officers who held municipal functions are veterans. Two of them, legionaries, both former *signiferi* in the legions *XV Apollinaris*, respectively *V Macedonica*, became decurions in the cities where they settled after discharge: Sarmizegetusa, respectively Napoca. Two veterans, former centurions in the auxiliary troops *coh. I Vindelicorum*, respectively *n. Palmyrenorum Porolissensium*, become decurions, one of them even *IIvir* also in the cities where they settle down, Sarmizegetusa, respectively Napoca. The same situation is true for a former *custos armorum* from the above-mentioned cohort who becomes decurion at Sarmizegetusa.

For the whole province we encounter 25 veterans occupying municipal functions, representing 9,22% of the total number of those attested, a rather high percentage that is comparable to the situation from other provinces from the Danube and Rhine.[418]

---

[415] Such an analysis for Lower Germany in Haensch 2001.
[416] Ardevan 1991, 169-171; Ardevan 1994, 203.
[417] Ardevan 1987, 121; Ardevan 1991, 169.
[418] Ardevan 1991, 170; Ardevan 1994, 204.

	Name	Rank	Unit	Honor	Place
1	C. Iulius Valerius	*veteranus ex beneficiarius consularis*	*XIII Gemina*	*decurio* *II vir*	Sarmizegetusa
2	C. Iulius Frontonianus	*veteranus ex beneficiarius consularis*	*V Macedonica*	*decurio*	Apulum
3	Valerius Valentinus	*beneficiarius consularis*	*XIII Gemina*	*aedilis*	Napoca
4	L. Valerius Rufus	*veteranus ex beneficiarius consularis*	*XIII Gemina*	*decurio* *quaestor* *II vir*	Sarmizegetusa
5	C. Iulius Fronto	*scriniarius praefectorum praetorio*	*coh. I praetoria*	*decurio*	Sarmizegetusa
6	C. Iulius Carus	*frumentarius*	*XIII Gemina*	*decurio*	Sarmizegetusa
7	Ulpius Masculinus	*veteranus ex signifero*		*decurio*	Napoca
8	Q. Manlius Verus	*veteranus ex signifero*	*XV Apollinaris*	*decurio*	Sarmizegetusa
9	M. Valerius Longinus	*medicus*	*VII Claudia*	*orn dec*	Drobeta
10	P. Aelius Theimes	*veteranus ex centurione*	*coh I Vindelic*	*decurio*	Sarmizegetusa
11	Laecanius Paulinus	*veteranus ex custode armorum*	*coh I Vindelic*	*decurio*	Sarmizegetusa
12	Aelius Thema[rsa?]	*veteranus ex centurione*	*n Palmyr Porol*	*decurio*	Napoca

Table I.8. Civil functions of the *immunes* and *principales* in Dacia.

Part II. CENTURIONS

II.1. General facts

Centurions are considered to be part of the plebs, as it was understood under the Principate, members of the *gregarii*, just like the *milites*. They were also granted *mission* but they were not considered ordinary veterans. Even so, centurions *qui ordine ducunt* are different from the *milites*, *qui ordine merent*.[419]

During the Principate the position of centurion was the highest rank of *militia*, opportunities for further promotion appearing only towards the end of the Republic and being established at the beginning of the Principate.[420] The importance of centurions is well known; they are the only category of professional officers in the army.[421] The most responsible officers of the legion,[422] they take care of all administrative matters concerning their *centuriae*, the transmission and execution of orders, the command of soldiers in battle and discipline and training. Superior officers, sometimes less experienced, could use the plentiful experience of centurions. The centurions can command detachments of soldiers, legionaries or auxiliaries, or they can act as diplomats. Due to their education, origin and status centurions are not associated to soldiers, forming a group of professional officers, loyal to the emperor.[423] Although they are the foundation of the army,[424] neither generals are amateurs, their abilities being known. They have a whole staff, a *consilium*, and a guard unit at their disposal where they often co-opt the centurions.[425]

Little information has passed down to us from the time of the Principate regarding their place and role in the battle. The situation must be associated to the one described by Caesar, although it cannot be completely identical. It is likely that by now the survival of the two parts of the Republican maniple – although it had lost its entire tactical role - was an accepted fact, once the armament and new battle line disposition were uniform. According to inscriptions dated during the Empire, the soldiers of a maniple were divided in *antesignani* and *postsignani*, probably the former being led by the *centurio prior* while the latter by the *centurio posterior.* The depth of the formation thus obtained was optimal for stopping cavalry charges and for narrow spaces and it is probably used starting with the end of the 2nd century AD. In Caesar's army the centurions fight in the first line of battle[426] and the multitude of decorations and the heavy losses in their numbers under the Principate prove their physical and tactical involvement in the fight.[427] On the battlefield the centurion is not very visible, being helped by the *optio* thus the supposed command of a whole cohort must be excluded. The centurion must participate, coordinate and evaluate trainings. He inspects the soldiers in the morning, reports to the tribunes and he receives the orders and the daily password.[428] They assign chores in the *centuria* and the administration of daily business, they enforce rules and discipline and they can strike any citizen with the *vitis*. In affairs involving the legion and the civilian inhabitants from its territory – *consistentes* – they serve as intermediaries and the *primipili* as a kind of representatives.[429]

The authority of the centurion is purely disciplinary. *Disciplina militaris*, a form of defining Roman military culture, becomes at the beginning of the Principate *disciplina Augusti.* It is enforced in the territory by his legates *pro praetore* who charge centurions with this task.[430] They exercise it with a rod made from grapevine – *vitis*; with that they impose their will or they punish. Lucilius' outburst,[431] which broke a rod while beating a soldier and then asked for another proves that this *vitis* could be considered a veritable weapon. The symbolic authority of the *vitis* in mentioned in legal acts:[432] if a rebellious soldier breaks the *vitis* he is to be put to death.[433] From the early beginnings of the Roman army the rod was a symbol of authority, coercion and military discipline. It was probably the inspiration for the *vitis* that is depicted on monument as a straight rod, between 0,45-1 m in length. *Vitis* can allow all by itself the recognition of a centurion, in graphical representations superior officers being depicted with the *virga.* The oldest monument depicting a centurion is the stele of Minucius from Padova,[434] from Caesar's time.[435]

The centurion also stands out because of his uniform and equipment. Funerary stones show him wearing a *lorica hamata* or *anatomica*,[436] the sword on the left side and a mantle coming down elegantly.[437] As for the helmet, it was noticed that it did not have a standard model.[438] The only singular aspect is the *crista transversa*, although this might be an artistic convention, in any case a sign of

[419] Domaszewski 1908, 80. For the discussion about the terms *ordinarius* and *ordinatus*, see Gilliam 1940, 127-144.
[420] Richier 2004, 445.
[421] Dabrowa 1993, 100.
[422] Webster 1969, 118.
[423] Campbell 1994, 47.
[424] Junkelmann 1986, 110.
[425] Josephus, *BJ* 6.262; Arrian, *Ektaxis* 22. Goldsworthy 1996, 123-124.
[426] *BG* 5.44, 6.40.7, 7.47.7, *BC* 1.46.4, 3.53.5.
[427] Richier 2004, 536.
[428] Josephus, *BJ* 3.86-88.
[429] Richier 2004, 537-538.
[430] Tacitus, *Agricola* 28 says that the tribunes who were leading the legion charge centurions with enforcing discipline. Speidel 2009, 29, 112.
[431] *Cedo alteram!* Tacitus, *Ann.* 1.23.
[432] *Digesta* 49.16.13.4. Phang 2008, 138.
[433] Brand 1968, 80.
[434] This centurion, probably from the legion *Martia,* disappeared sometime between 44-42 BC, does not wear armor or helmet, only a mantle over a tunic, a *gladius* and a *pugio*, the only sign of his rank being the *vitis*. Keppie 1991, 117.
[435] Feugère 1995, 114-115.
[436] In the case of M. Caelius-dead at the Teutoburg Forest, Feugère 1995, 116.
[437] Centurions wore red tunics in order to stand out on the battlefield (*HA*, *Claud.* 14.5, *Aurel.* 13.3). Phang 2008, 83.
[438] Feugère 1995, 116.

recognition on the battlefield for the soldiers.[439] The centurions own a horse, they march mounted and they wear their decorations on the equipment.[440]

Unlike the moral and intrinsic authority of superior officers, that of the centurion is more physical, more personal and obviously more efficient.[441] His image is that of the accomplished officer.[442] Soldiers were used to obeying their centurion, with whom they had the most social contact and who could punish them. Centurions and officers followed the code of manhood, governed by *virtus*, looking and behaving as manly as possible in order to legitimate their command. Centurions inspect the building of the fort.[443] In the presence of a centurion or superior officer a soldier or an underling had to stand up.[444] The interdiction of legitimate marriage for all soldiers, including centurions was effective from Augustus until AD 197, when Severus lifted it.[445] Until Severus only the centurion had the right to physically punish soldiers; afterwards this is also permitted to *speculatores, quaestionarii* and *beneficiarii.*[446] Severe insubordinations, such as assaulting a superior officer or refusing corporal punishment were punished by death.[447] In the case of punishment for unbecoming behavior centurions enjoy more favorable treatment as compared to soldiers, who are usually decimated, the former not being used to physical punishments.[448] Besides their payment, considerably greater than that of the soldiers, centurions enjoy unofficial "prizes". Soldiers obtain money to bribe their centurions to be exempted from chores. Centurions grant leaves for soldiers – *commeatus*, in return obtaining favors.[449]

Tactical discussions or even speeches before the battles were discussed in the war council, where some centurions were allowed but no soldiers, who considered their officers as somewhat responsible for their fate. In case of disagreement between commanders of equal rank the vote and opinion of centurions was considered in decision-making. The body of officers was considered separate from that of the soldiers and sometimes the latter rebel even against centurions.[450] In a case of total rupture the soldiers retreated in one of their forts, chose centurions and made their own vows, thus completely reorganizing themselves.[451] In any case, any opinions or problems the soldiers might have they reach the commanders only through the centurions and tribunes.[452] The rank of centurion was very prestigious, as proven by knights willing to serve as centurions. It was a giant leap for a soldier, whose income, but also expenses, increased dramatically. When Pliny the Younger made one of his clients, Metilius Crispus, centurion he offered him 4,000 sesterces for equipment.[453] Centurions had several slaves as personal property.[454] Access and promotion to the rank of centurion were only partly based on merits. The most important was patronage, as in the examples of Pliny, with Metilius Crispus or Nymphidius Lupus, for whom he practically writes a recommendation letter for Trajan.[455]

Each centurion has his personal tent, a considerable quantity of luggage and luxury items, carried by at least one mule.[456] Centurions were entitled several food rations, that is why they were written down on a separate registry.[457] Inside his own home the centurion eats separately, probably with the *optio* and *signifer*.[458]

His rank is awarded for life. The position of centurion was obtained by men belonging to military elite[459] to a privileged class. This ranks awards zealous soldiers who, after gaining command, dedicate themselves to military discipline, through the *vitis*. A tool of command, the centurion is also an example for the soldier. This intermediary position offers a unique character for this rank, making him the 'most' professional officer in the Roman army.[460]

[439] Vegetius, *Epitoma* 2.13-16. Even though such a representation is visible on the stele of T. Calidius Severus, CIL III 11213, Carnuntum: *lorica hamata, ocreae* and a helmet with a *crista transversa*. Bishop, Coulston 2006, 96, fig. 52/2. Another such representation is that of centurion L. Valerius Seranus, from Kovin (Serbia), IDR III/1 1, even though only the drawing of the stone survived.

[440] Webster 1969, 132; Feugère 1995, 117.

[441] Brand 1968, 81.

[442] Although in Italian public opinion from the time of the Civil Wars, soldiers and centurions are regarding as killing machines of incredible cruelty – Appian, *BC* 4.4.17. Phang 2008, 273.

[443] Vegetius, *Epit.* 3.8.13.

[444] As understandable from the recognition and assassination of Pompey (Appian, *BC* 2.12.84). It is not certain whether soldiers were allowed to speak first, in Dio 69.19.1 the centurion Similis saw Trajan before the prefects and remarked this as unusual. Appian, *BC* 3.12.88, considers that it is improper for centurions to address senators directly, like they did when they asked the office of consul for Octavian. Phang 2008, 18, 34, 68, 86-88.

[445] Phang 2008, 92. Hoffmann 1995, 110 claims the status of legionary centurions must have been equivalent to that of the *praefectus cohortis*, whose rank he equals and thus he might be allowed to marry.

[446] *Digesta* 49.16.13.4

[447] *Digesta* 49.16.6.1 – if the soldier breaks the centurion's *vitis* or touches it.

[448] Phang 2008, 125.

[449] Phang 2008, 176, 211-212. Details about the grant of *commeatus* in Wesch-Klein 2000.

[450] Tacitus, *Hist.* 1.18, Suetonius, *Otho* 1.2, Dio 79.32.3.

[451] Appian, *BH* 34, Caesar, *BC* 1.20.

[452] Caesar, *BG* 1.43, 7.17, *BC* 1.64. MacMullen 1984, 453-455.

[453] *Ep.* 6.25.2.

[454] CIL III 8143 – nine slaves, Mathew's Gospel 8.5-13. CIL XIII 8684 indicates that one centurion had two slaves as personal property that apparently died with him in the Teutoburg Forest. Roth 1999, 103; Phang 2008, 235.

[455] *Ep.* 10.87. Juvenal also speaks about the importance of recommendation letters in the army in *Sat.* 16.4-6. Goldsworthy 1996, 31-32.

[456] Goldsworthy 1996, 290.

[457] Fink 1971, 10; Roth 1999, 22.

[458] Roth 1999, 58.

[459] Tacitus, *Ann.* 1.17.1

[460] Le Roux 1982, 292.

II.1.1. Centurions' origin and entry to the rank

Academics have been writing on the origin of centurions for more than one century, starting from two different positions. The creator of the first opinion, Domaszewski, started from the principle of keeping intact the Roman-Italian component of legionary centurions in order to obtain tactical and loyal unity of the army, in his opinion until the 3rd century AD.[461] The other opinion is based on the analysis of epigraphy of centurions outside the confines of any theory as done by the British school that reached different conclusions.

The one who induces a complete reform of old theories is E. Birley who proposes a whole new direction in the study of this rank. Taking into account that most centurions are in fact simple names, written on tombstones or lists of centurions, these names, especially if they are Roman, cannot indicate clearly the origin of the officer.[462] Obviously those with Roman names may not be Italians because the Roman system of awarding citizenship implies the dissemination of Roman names. Based on this aspect, lists of centurions that contain, after the name, the *origo* are useful.[463] The dating of these inscriptions is difficult, except for the ones already dated.[464] When noticing the multitude of centurions who do not have Italian origin, Domaszewski's theory is not verified.[465]

B. Dobson, this 'student' of E. Birley, follows in his professor's footsteps enouncing quickly the evolutions before the publishing of the critical edition of *Rangordnung* and also opening the way for prosopographic studies dedicated to each legion, starting with the second half of the past century. At the beginning of the Principate, the Italians dominated the legions, but rapidly their numbers begin to dwindle in favor of colonies already Romanized, from Narbonensis, Hispania and Africa. Military traditions that spread in the provincial colonies, but Italian as far as population is concerned, will contribute strongly to groups of soldiers but also centurions and officers of equestrian rank. For example, we do not encounter Africans in Hispanic legions, indicating a preference for local recruitment. Starting with Hadrian the Italians no longer volunteer to serve in the legions, preferring praetorian cohorts and until the end of the 2nd century all legions received recruits from colonies of veterans of the provinces and also from the men living in the *canabae.* Once veterans settled down in the area were they served, military families are created, several specialized generations, which will offer the majority of recruits in the 2nd century, for the whole frontier army, as well as the majority of centurions and officers of equestrian ranks.[466] Concerning the prosopographic study of several legions, the efforts confirmed what E. Birley proved almost half a century ago: the gradual provincialisation of legionary and implicitly centurion recruitment.[467]

Reaching the centurionate. Although there are several variations on the same theme, most academics agreed concerning the sources of legionary centurionate. The main principle divides future centurions in three categories: former legionary *principales*, former praetorian *principales* and directly appointed centurions. A closer look might reveal several subcategories in between these categories.

B. Dobson and D. J. Breeze make the most detailed and complete description of sources of legionary centurions, while trying to establish exactly the connections between the cohorts from Rome and legionary centurions. E. Birley made a previous attempt.[468]

They accept the classification in three categories[469] for the sources of centurions, claim that all three classes enjoy the same opportunities,[470] but they establish several subcategories, depending of the options if the future centurions. Thus the options can be grouped in six branches:

A(1) – legionnaires, only towards legionary centurionate, until Severus, following 15-20 years of service;[471]
B(2a) – praetorian *principales*, towards legionary, but not praetorian, centurionate;[472]
B(2b) – *evocati*, towards legionary, but not praetorian, centurionate;[473]
B(3) – *evocati*, towards centurionate in Rome and then in the legions;[474]

[461] Domaszewski 1908, 83-90; Dessau 1910, 2-24.
[462] Birley 1941, 190.
[463] CIL VIII 18084.
[464] CIL VIII 18065, Lambaesis, a. 162; III 6580, Alexandria, a. 194; CIL XVI App. 13, Palestina, a. 150.
[465] Birley 1941, 191-198. Another essential contribution is that of G. Forni who centralized the origins of legionaries from the Empire in two books published 20 years apart and although he did not necessarily focused on the origins of centurions, sometimes even excluding them from his statistics, his study confirms what was already known: the theories regarding the increasing provincial character of the legions starting with the 1st century but especially from the 2nd century onwards. Forni 1953, especially 65-75 and Forni 1974, 380-390.

[466] Dobson, Mann 1973, 192-197. About the lack of British centurions see Dessau 1912 and Birley 1979, 76-77.
[467] P. Le Roux discusses the situation of centurions from Hispania. Le Roux, 1982, 294-299. Similar situations in Le Bohec 1989, 184; Christol 1994, 181-187, for the legion *III Augusta*, in Traverso 2000 for *II Augusta* and in Dabrowa 1993 102-103 for *X Fretensis*. To all of this we must add the contributions for each legion published together in the papers of a conference on this subject by Y. Le Bohec, as well as more recent contributions. Le Bohec 2000, Mosser 2003, Malone 2006.
[468] Birley 1941, 199-203.
[469] Accepted by most specialists: Le Roux 1982, Campbell 1984, Le Bohec 1989, Dabrowa 1993, Hoffmann 2005 *etc.*
[470] Dobson, Breeze 1969, 100.
[471] A(1) – do not offer details about the career before becoming centurions; they might serve on different centurion posts in the same legion; a petty officer is eligible for promotion at any time but promotion comes after 15-20 years of service (at the age of 35-40) in the same legion; they do not really have a chance to become *primipili* but this might be so because *primipili* do not present their whole careers.
[472] B(2a) – *cornicularii* of the prefects from Rome become only centurions in legions, directly, not in praetorian cohorts and in the legion they occupy directly a post within the *primi ordines.*
[473] B(2b) – *evocati* directly to legionary centurions do not present in detail their career prior to evocation; usually they are *urbaniciani* prior to becoming praetorians; recalling involves a service, not just a step to becoming centurion.

B(4) – centurions from Rome, without serving in the legions, but who can become *primipili* directly (late cases) ;[475]
C(5) – directly appointed, only legionary centurions;
C(6) – directly appointed, served as centurions in legions and in Rome.[476]

In all cases a career after a post of *primus pilus* is a prize, not a privilege awarded to a certain class.[477] The tables proposed by Dobson and Breeze highlight the disproportionality between Italians and provincials, between directly appointed centurions and *primipilares*, when it comes to setting up inscriptions. A preference of a certain category of centurions for the post of *primus pilus* cannot be observed. To conclude, the post of legionary centurion was mainly filled with soldiers from the legions; the *evocati* are granted the rank of legionary centurion as a reward, they become *primi ordines*, not necessarily *primipili*; their proportion and that of those directly appointed is small. The posts of centurions in Rome were mostly filled with *evocati* and few directly appointed centurions. The post of *primus pilus* is reached in about 50% of the careers of legionary centurions, but most remain on the post of *praefectus castrorum*. Most of the centurions in Rome become *primipili*, in order to achieve this they serve on a post of legionary centurion; those who lack this post are only very early and under Marcus Aurelius. Directly appointed centurions do not seem to follow any rules. They hope for help and patronage in order to be promoted towards *primipilus* or to obtain a posting as centurion in Rome and then *primipilus*. The primpilate was not automatically granted for any category, least of all a career following it.[478]

One could become legionary centurion following a mandate from the governor and centurion in Rome following a mandate from the respective prefect. Most seem to achieve this rank at around 30-35 years of age and a service of at least 20 years could be necessary before they can become *primipili*, during which many fail.[479]

A conclusion that can be obtained from these statistics is that for the *ex caligati* in most cases the post of centurion was the culmination of a beautiful career that could include service in the auxiliary or praetorian troops.[480] The appointment of centurions was one of the emperor's attributes, through his secretary, *ab epistulis*,[481] although the governor had a certain say in filling up empty positions.[482] One of the methods used for promotion was the consultation of the soldiers – *suffragium*. In any case, an emotional motivation was involved when selecting those serving in the *officium consularis* for promotion. In the cases where we know the career as *principales*, it can be noticed that centurions were promoted inside their units or inside the provincial army they originated from.[483]

Directly appointed centurions belong to the provincial elite, they may or may not be knights or even sons of centurions; anyway their possibilities depend on the quality of their origin and their father's. As far as knights are concerned, most become centurions at the beginning of their careers, but there are also those who join the army after starting an equestrian career, following one, two or even three *militiae*, because the opportunities of a equestrian career of a *primipilaris* are more financially rewarding. Many times previously held equestrian commands offer a bonus in the hierarchy of his appointment as centurion but anyway, in this case, connections were very important. They become younger centurions, around the age of 30 and thus have more time to promote.[484] What happens to their status as knights while serving as centurions? Although it has been proven countless times that they do not have to give up being members in the *ordo equester*, more recently, it has been stated that the post of centurion is part of the *militia*, being its highest rank and knights willing to become centurions had to renounce membership to their order while serving. From the time of Severus, when everything changes completely, there seems to no longer be a barrier between the *militia* and the *ordo equester*. All knights had to occupy a post of legionary centurion for the first time because they could not become praetorian centurions directly.[485]

Case studies focused on centurions from certain legions can complete the general picture.[486] The unique situation of L. Maximius Gaetulicus, who in AD 184, when he was *primipilus leg. I Italicae*, at Novae, fulfilled a promise made 57 years before, when he was *tiro aput leg. XX Valeriae victricis*[487] is an exception; not all young recruits could hope for this but it proves that the hope of becoming *primuspilus* was the main stimulus for those seeking to reach the post of centurion and who stay in service much longer that the statutory period.[488]

---

[474] B(3) – *evocati* go for a centurion position in Rome then in the legions, they occupy the positions of *trecenarius* or *princeps castrorum*; *trecenarius* seems to be mandatory in order to have a chance to become *primus pilus*; they serve only once as legionary centurions prior to becoming *primus pilus*.
[475] B(4) – those who serve as centurions only in Rome then move on directly to *primus pilus* seem to be later cases, quite unclear.
[476] C(5 and 6) – those directly appointed, become legionary centurions then some become centurions in Rome; probably the appointment was due to the governor's favor, that is why they appear mostly in legions, imperial favor being needed for an eventual transfer to Rome. The rank of *primus pilus* is common in this category but this is not the most favored one. Dobson, Breeze 1969, 101-102.
[477] Dobson, Breeze 1969, 103-111.
[478] Dobson, Breeze 1969, 114-117. Following that, Dobson concludes that legionnaries had a clear advantage when seeking to become centurions. Dobson 1974, 403-407.
[479] Dobson 1974, 406-407.
[480] Richier 2004, 445.
[481] Statius, *Silv.* 5.1.94, Juvenal, *Sat.* 14.195, Florus, 108.
[482] Tacitus, *Ann.* 2.55, *Hist.* 1.52, 3.44.
[483] Richier 2004, 451.
[484] Richier 2004, 460-462.
[485] Richier 2004, 463.
[486] For Hispania Le Roux 1982, 300-301, for the Rhine area Richier 2004, 508-514, for Pannonia Mosser 1003, 116-117
[487] He was previously known as centurion in the *XX Valeria victrix* from RIB 1725 and RIB 2120.
[488] In any case there are only two other centurions who serve for a longer period of time: CIL III 11031 becomes *primipilus* after 58 years

In more recent inscriptions we see more and more centurions enlisted in the army while very young, under 30 and some even under 20 years, in the *VII Gemina, III Augusta*,[489] Sex. Pilonius Modestus,[490] Ti. Claudius Fatalis,[491] Petronius Fortunatus,[492] so we cannot state that centurions were not selected young. Compared to them, a series of centurions who served for a long time as soldiers and/or *evocati* enlisted quickly in the army but became centurions only around the age of 40.[493]

Probably soldiers chosen for promotion to centurion stand out because of their abilities, competence and education. Civilians who had possibilities and relations also tried to become centurions. Pertinax, although he started as a petty functionary, tried to become centurion using his father's connections.[494] Although the centurion recommended by Pliny receives money for his equipment from him, he was probably very well paid because he had slaves.[495] Dio,[496] while opposing to former soldiers being elected in the senatorial order, does not oppose those knights who had served as centurions. Juvenal[497] relates how a greedy father pressures his son and sends him with a written petition to become centurion and get wealthy. The responsibility and importance of centurions can be observed from their status and payment. The emperor assumes that centurions, being distinguished because of their status, could appreciate the possibilities for promotion offered by a professional military career. Thus, a closed group of elite, professional officers was created, loyal to the emperor in any circumstance. It has been suggested that emperor was in charge of promoting centurions but it is unlikely that he had the time for all this, although such promotions are registered in Rome.[498]

Normally, promotions to centurion were suggested by the next higher-ranking officer, then sent up the chain-of-command and eventually approved. The emperor interfered probably in special situations.[499] Soldiers who committed extraordinary acts of bravery are marked for promotion by the emperor himself,[500] but in general it must have been quite unusual for the emperor to involve in the soldiers' military careers.[501]

and another also becomes centurion but after 61 years of service. Birley 1989, 114.

489 CIL VIII 2869.

490 CIL III 1480.

491 AE 1939, 157.

492 CIL VIII 217.

493 Birley 1989, 115.

494 *HA, Pert.* 1.4-5.

495 Campbell 1984, 103.

496 52.25.7.

497 14.133-139.

498 Campbell 1984, 103-104.

499 Such an example is the career of Ti. Claudius Maximus, the captor of Deceballus, who mentions on his tombstone he was promoted several times by Trajan himself.

500 ILS 2666, 7178.

501 Campbell 1984, 105-108.

II.1.2. Centurion's hierarchy

Centurion's number and titles. The study of the organization and functioning of a legion might have reached a certain consensus, especially concerning the division in tactical subunits. Also, the number of these subunits, the ten legionary cohorts, are generally accepted by all scientists. After the legion evolved from its organization on maniples to that on cohorts, its division has been done in ten cohorts, each containing six *centuriae*, for a total of 60 centurions.[502] This organization survived also during the beginning of the Principate, as revealed by literary sources.[503] Thus we move on to another subject that does not enjoy the same general consensus: the number of centurions. Logic leads to a number of 60, given the division of the legion. The 60 centurions in a legion are divided in several ranks and categories.[504] Still, the problem appears in the first cohort. As discussed on other occasions, it was reorganized at some point, its number of soldiers doubled and one of its centurions disappeared.[505]

Typical in inscriptions of centurions is the mention *centurio* and the legion where he served (*centurio legionis*), without any other details concerning his post, place or rank.[506] Anyway, this is the most encountered reference to centurions in epigraphy.[507] Still, it is considered a certain fact the number of six centurions in the other nine cohorts. Each carry specific titles, as a form of recognition, placement or for establishing the hierarchy, The titles of the six centurions from one of the cohorts II-X are classified based on the deployment in the battle line – *triarii (=pili), principes, hastati*, cohort – *II-X* and position in the maniple – *prior, posterior*,[508] as it follows:

502 Domaszewski 1908, 93, presuming two *primi pili*; Parker 1928, 196.

503 Tacitus, *Annales* 1.32.3.

504 Marquardt 1876, 357; Domaszewski 1908, 93; Parker 1928, 196; Webster 1969, 118; Le Bohec 1994, 43.

505 The problem regarding the number of centurions in the first cohort appeared in the 19th century and it is still valid, with two main variants, 5 or 6. Literary sources are not completely clear. Polybios, 6.21.7, indicates 30 maniples with 2 centurions each, thus 60. Aulus Gellius 16.4.6 describes the situation from the time of Augustus, similar to the one described by Polybios, except he refers to the number of *centuriae*. Tacitus, *Ann.* 1.30.2, speaks about the number of centurions that is 60. The epigraphic finds unfortunately complicate the situation, CIL VIII 18065 and XIII 6801 give a chaotic number of centurions. The lists of *optiones*, CIL VIII 2554 from Lambaesis, indicates 60, but we cannot presume that is also the number of centurions while a second one, CIL VIII 2555, mentions only the five *optiones* of the centurions from the first cohort, in the order of the centurions they serve – *primus pilus, princeps, hastatus, princeps posterior, hastatus posterior*, a situation that is identical in the second part of the inscription, dated under Valerian. The situation is confirmed by AE 1969/70, 633, a list of veterans discharged in AD 156-157, at Nikopolis *ad Alexandriam*, grouped according to *centuriae*; they are five in the first cohort, starting with the *primus pilus*. The double number of the first cohort mentioned by Hyginus, 3.4 and Vegetius 2.6, 2.8, is also confirmed by CIL III 6178 and 14507, both of them lists of veterans indicating a double number of discharged soldiers from the first cohort. The archaeology of legionary fortifications confirms this statement. Parker 1928, 197-199; Keppie 1984, 151; Campbell 1994, 46; Le Bohec 1994, 43.

506 Wegeleben 1913, 23.

507 Richier 2004, 464.

508 Marquardt 1876, 357.

pilus prior,
princeps prior,
hastatus prior,
pilus posterior,
princeps posterior,
hastatus posterior.[509]

For the first cohort:
primus pilus,
primus pilus (posterior),[510]
princeps,
hastatus,
princeps posterior,
hastatus posterior.

As previously stated, the most encountered title in epigraphy is *centurio legionis*, without any other mentions, quite complicating the situation as far as hierarchy is concerned. There are variations in the names, for example for the first cohort, where the centurions *priores* are called simply *princeps* and *hastatus*, without any number or mention to their position in the maniple. As expected, the sixth centurion of the first cohort has long been searched for, unfortunately without success or with limited one. In the last years the subject of the names of *centuriae* and of the titles of centurions has been reevaluated, due to tactical changes in the functioning of the legion that can be observed from the end of the 2nd century but especially in the 3rd century. Luckily Dacia helps to clarify these evolutions. CIL III 8031 mentions the fortification works around Romula made by a detachment of soldiers from the legion *VII Claudia* who leave as evidence a plaque, dedicated to the emperors Philippi.[511] Three other inscriptions found in the wall name other *centuriae* that took part in tis construction.[512] The practice of legionary detachments to name their *centuria* according to cohort and rank and not according to the name of the centurion is often encountered in this period, in the present case both *centuriae* of *hastati* from the VIIIth cohort are involved.[513]

M.P. Speidel continues the discussion in a series of works published in a time span of over 20 years. He starts from the titles of the centurions that are in the following order – according to rank: *pilus prior, princeps prior, hastatus prior, pilus posterior, princeps posterior* and *hastatus posterior*. Each title also includes the cohort. During the Principate the word '*cohors*' disappears from the centurion's title, only the numeral and the post remaining, just like in the army of the Republic: *III (tertius) hastatus prior*.[514] Because the word '*cohors'* is not mentioned in the inscription there is no reason to place it there, especially since the numeral written in front of the post is ordinal. Anyway, in this manner we have epigraphic evidence for half of the posts in the legion.[515]

Besides, the *centuria*, though initially named after the one who was centurion at that time, starts to be named impersonally starting from the 2nd century. The evolutions follow several systems: the system *II Traiana* with or without mention of the centurion: *coh(ortis) II (secundae) hast(ati) pr(ioris)*, in Egypt and Britannia, or that mentions the cohort, then the sign of the *centuria* '>' and the name and rank of the centurion. The second system, *II Parthica*, appears sometime at the end of the 2nd century and comes closer to the idiomatic titles of centurions: > *(centuria) IX (noni) prin(cipis) posteri(oris)*.[516] This second variant appears mostly on monuments belonging to legionnaries from vexillations and expeditionary armies. A particular situation is encountered in Mogontiacum[517] where the abbreviation is even more severe: *genio > II Γ*. The depersonalization of these titles appeared most likely in order to more easily distinguish between the vexillations from several legions, brought together in expeditionary bodies from the 2nd-3rd centuries.[518] An essential contribution to this problem was offered by a series of inscriptions that present a number of graphical that might abbreviate the titles of legionary centurions. M.P. Speidel, twice, and J.C. Mann have dealt with this problem in the end reaching similar conclusions. In his first work M.P. Speidel recognizes these signs on a number of monuments from different places in the Empire.[519]

The form and function of the centurial signs indicate they are part of a general design of the shape of the cohort on the battlefield. The best arrangement of signs is exemplified below.[520]

[509] Domaszewski 1908, 93.
[510] This second *primus pilus* has been identified with other titles – *primus pilus bis, triarius ordo, centurio veteranorum, centurio supernumerarius etc.* More recently Sarnowski 1993a, 209, considers the second as no longer serving in that post but also paying for the monument, just like Domaszewski. Other unclear attempts in Keppie 1993, especially 11-16; Richier 2004, 465-467.
[511] Speidel 1978a, 121.
[512] AE 1940, 13 and 14, CIL III 8047.
[513] Speidel 1978a, 119-120.
[514] Speidel 1983, 43-44, although several academics, including Mommsen, insist that the cohort continues to be mentioned.
[515] Speidel 1983, 45.
[516] CIL III 187; Speidel 1983, 47-48.
[517] CIL XIII 6681, from AD 230.
[518] In Latin these titles are called *ordines*, Tacitus, *Annales* 1.44. Speidel 1983, 49, 61. An earlier parallel might be the depersonalization and standardization of the names of auxiliary troops from the time of Tiberius: they are no longer named after their commanders. The titles of *centuriae* and the mention of the cohort prove that these subdivisions were used both tactically and administratively during the Principate. Speidel 1983, 50.
[519] CIL XIII 6681 and 6683, Mainz, a marble tablet from Alexandria, IDR II 326, CIL VIII 18065, Lambaesis, and one inedited inscription from the fortification from Mainz. Speidel 1986a, 321-330. Contributions also in Mann 1997, 295-297.
[520] Speidel 2005, 290. The most recent contribution in this direction does not change the significations that have been proven so far with great difficulty but adds several examples to the list of inscriptions that contain these signs. A recent discovery from Apulum is included, but the authors who published it did not recognize these signs, ignoring them completely. The succession of the three signs, perfectly readable in the epigraph, is: *princeps prior, hastatus prior, hastatus posterior*. Moga, Drâmbărean 2007, 176-177; Faure 2008, 298.

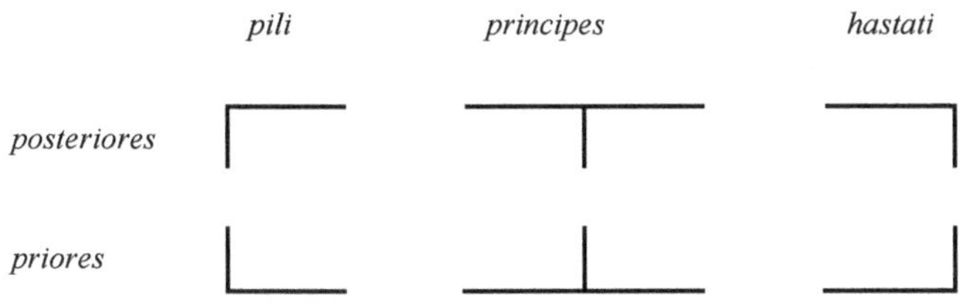

The most troublesome problem of the legionary centurionate is the hierarchy and implicitly the promotion system. After more than 100 years of research a general consensus still has not been agreed upon; even now there are still two directions of study and understanding of this phenomenon. The two directions were both theorized in German academic environment and they are totally opposite. In the study of the *Rangordnung* they are known by the names of their authors: A. von Domaszewski and Th. Wegeleben.

Domaszewski's theory. It must be stated that there were other German academics before Domaszewski who tackled the problem of the legionary centurionate, in the second half of the 19th century, the most important being Marquardt.[521] Other German scientists dealing with this are Mommsen, Müller and Lange; they all propose similar theories or small variations of Marquardt's theory.[522] In a way Domaszewski is tributary to these previous attempts but the theory he created is much more detailed and better proven.

First of all Domaszewski accepts Mommsen's opinion who stated that the rank of the centurion depends on the number of the cohort. The lowest ranking are those in *cohors X*, the highest-ranking are those in *cohors I*. The hierarchy of those in the first cohort is discerned from inscriptions:[523] *primus pilus, princeps, hastatus, princeps posterior,* and *hastatus posterior*. They are the commanders of the *centuriae* of *cohors I*. Promotion was from *princeps* to *primus pilus*.[524] This promotion is attested in several inscriptions and is due to the principle stating that always before reaching the position of *primus pilus* a post of legionary centurion must be held.[525] Promotion from *hastatus* to *princeps* is also mentioned in inscriptions.[526] These three ranks form a closed group, the highest-ranking one and admittance to this group is obtained by promoting to *hastatus*.[527] The centurions *posteriores* from the first cohort carry their titles since the beginning of the Principate. Inscriptions attest a group comprised of *princeps posterior* and *hastatus posterior* but they do not mention any *pilus posterior* in *cohors I*. In other words the division of the first cohort in five *centuriae* must date from the beginning of the Principate. This division can take place on the basis of the tactical change of the legion, because of Marius' reform involving maniples. All other nine cohorts have six *centuriae* and centurions are mentioned (including *pili posteriores*).[528]

According to the titles that these centurions had, he presents the ranking order: *primus pilus, primus pilus (posterior?), princeps, hastatus, princeps posterior, hastatus posterior (cohors I), pilus prior, princeps prior, hastatus prior, pilus posterior, princeps posterior, hastatus posterior (cohortes II-X)*. According to Polybios[529] promotion could take place only inside the same age group. Centurions are divided in two categories: *priores*, chosen first and *posteriores*, selected afterwards. The difference was that the *prior* commanded the right side of the maniple that in Antiquity it had to bore the grunt of the battle. Hierarchy is then established according to the number of the maniple, from I to X. Promotion took place inside the squadron of *priores* or *posteriors* from maniple X to maniple I (*X hastatus posterior – I hastatus posterior* etc.). A senior centurion could move from the *posteriores* to the *priores* and advance in this category.[530]

The centurions from the first maniple, from each age group are the highest ranking, named in the Polybian fashion *primi ordines*. When the organization based on cohorts was introduced, the first three maniples were gathered in *cohors I*, which also becomes first in the hierarchy thus *cohors X* becoming the last one.[531] Those from *cohors I* are *primi ordines* during the Principate also and now there is a clear difference between them and the rest of the centurions.[532] This is very important for the

[521] Marquardt 1876, 358-361. Starting with the presentation of the titles of centurions he claims that during the Republic a promotion system cannot be acknowledged because centurions are recruited each year and thus on random positions. For the period of the Empire he wrote a theory stating that *X hastatus posterior* is the lowest rank of centurion and his promotion involves him passing through all those posts in each cohort, then he joins the *principes posteriores* and so on until he finished all 30 posts of *posteriores*. Following, he joins the *priores* while respecting the same system until he reaches the position of *primus pilus*. At the beginning of the Principate the maniples are replaced with cohorts, the names of the centurions remaining the same, except that now the number signifies the number of the cohort. In this situation the promotion system changes, a centurion moving through all six posts in a cohort in the order *hastatus posterior – hastatus prior – princeps posterior – princeps prior – pilus posterior – pilus prior* and so on for each cohort. Another problem is the signification of the centurions *primi ordines* and their hypothetical role as commanders of each cohort. Marquadt's two options are *pili priores* of each cohort or the three *priores* of each cohort. He prefers the first one because he considers the cohort to be a closed military body with its own administration and command. Beside, just like the maniple is commanded by the centurion *prior* so is the cohort commanded by its first centurion. Thus advancement in the cohort does not allow for transfer in the next cohort on an inferior post, *hastatus* or even *posterior*.

[522] *apud* Wegeleben 1913, 11-14.

[523] CIL VIII 18072.

[524] Vegetius, *Epitoma* 2.8.

[525] CIL IX 4122, XI 5215, 5273. Domaszewski 1908, 90.

[526] CIL XI 5215.

[527] This group might include a *tribunus angusticlavius* who wanted to serve as centurion. cf. CIL X 5829=IDRE 90.

[528] Domaszewski 1908, 91.

[529] 6.24.2.

[530] Domaszeski 1908, 93.

[531] Caesar speaks about the *octavi ordines* as centurions of the *VIII* cohort in this manner.

[532] In inscriptions (CIL VIII 18042, 18065, XIII 6801) we find wordings like *primi ordines et centuriones*. More recently on a base of a statue for Severus dedicated by the *primi ordines et centuriones legionis I Italicae* in *principia* of the fortification from Novae; Sarnowski 1993a, 205-212..

promotion system of centurions during the Principate. The basis for promotion is the cohort which now becomes a standard. Rules can be deduced from CIL VI 3584 = IDRE I 3 and CIL III 7334, also showing that transfer from one legion to another is a promotion.[533]

To conclude, Domaszewski accepts two systems of promotion, one for the centurions *ex equite Romano* or those coming from the cohorts in Rome – *Staffelavancement*, on echelons, and another one for centurions *ex caliga* – *Stufenavancement*, on ranks. A centurion *ex caliga* could benefit from promotion on echelons if he showed bravery in combat. *Staffelavancement* means only 16 promotions, between cohorts, possibly also inside the same cohort if we consider the first cohort. His great contribution lies in explaining the terms *succedere*, *successione promoveri* and *promoveri*, meaning promoting from one rank to another, promoting from one cohort to another and promoting while skipping over ranks or cohorts while at the same time being transferred.[534]

In the same year that *Rangordnung* is published, G.H. Allen publishes his contribution to officer promotion in the Roman army but he moves very quickly over the promotion of centurions, without having something to say.[535] Shortly after this, M. Radin's study on the promotion of centurions in Caesar's army will be published. In his work he reconsiders the opinions of German academics from the 19th and the beginning of the 20th centuries, trying to find arguments for each of them but their only common place is promotion through all 60 posts of the legionary centurionate.[536] Also in the Anglo-Saxon academic environment, H.M.D. Parker tackles the promotion system of centurions in two papers, published two years apart, his contribution being integrated in the extensive study of Roman legions. He believes that there were two centurions with the same rank, differentiated only by one being *prior* and the other *posterior*, in the sense of seniority.[537] Parker's theory is in fact a combination of several others previous.

In the next period efforts to prove or augment Domaszewski's theory are absent, the tendencies going in a complete opposite manner. Still, the Austrian academic's contribution was very important and he practically opened a direction of study that will be active even after 100 years.[538]

Unfortunately not all contributions to the study of the idea of hierarchy that were published after Domaszewski's paper were a success, most of them succeeding in complicating things very much instead of explaining them. The same is not true for the other line of study concerning the hierarchy of centurions.

Wegebelen's theory. Apparently Th. Wegebelen must give some credit to a previous contribution of E. Desjardins, published in 1884. He starts from the idea that the names of centurions in inscriptions are made only by mentioning the *centurio* and the legion to which he belonged (*centurio legionis*), without any other details regarding the post, place or rank.[539] In any case this is the most often-encountered centurion title in epigraphy.[540] We must accept the fact that the silence of the sources regarding this detail suggests that it had no meaning whatsoever in Roman times, not even for the centurion's colleagues. No inscriptions mention in detail a career, not even those used by Domaszewski. Generally posts are not mentioned, indicating their lack of importance even in the case of specialized military personnel. There was no difference between two centurions from the cohorts II-X, neither between them nor in regard to the superior officers. Arguments: 1. The overwhelming number of inscriptions mentioning only *centurio*, as compared to those mentioning posts; 2. The lack of mentioned posts even in inscriptions where this would be expected or necessary; 3. We can never establish for certain more than one post held in the cohorts II-X; 4. The careers of *principales* record every post because there is a difference in rank between them, so too do the tribunes who were centurions or *primipili*.[541]

There were probably differences in seniority, length of service or in names (per cohort or maniple) but as far as promotion is concerned these are all irrelevant. Rank differences cannot be determined from literary sources either. Because most centurions are identical in rank it is only personal merit and military achievements that differentiate them. The longer they occupy these posts the more we can consider them unable to become *primus pilus*. Obviously we must differentiate between

[533] Domaszewski 1908, 94. Normal promotion is expressed by the terms *succesione promoveri*, found in both inscriptions. The system can be traced based on the promotion system of the *optiones* who are moved in a superior cohort but on the same post. Caesar defines cohorts (*BC* 3.54) in the same way Polybios defines age groups as echelons, thus *succesione promoveri* means moving into the next echelon, on the same post by skipping the five posts in between. Thus, the example from CIL VI 3584 is promoted on the same post, *hastatus posterior*, in the ten cohorts except each time he is moved in a different legion. Other similar examples show the advantage given to praetorians that from the position of *evocati*, are directly appointed centurions as *principes* or *pili* in the first cohorts, some becoming even *primi pili*. They are joined by knights who serve as centurions and who are advantaged when promoted on echelons (*Staffelavancement*). By contrast, a centurion promoted from legionary soldiers during the time of Septimius Severus will spend his entire military service as centurion or might be promoted if he showed exceptional bravery and skill on the battlefield. Only the three highest posts, *hastatus*, *princeps* and *primus pilus* are held consecutively (*Stufenavancement*). Promotions in ehcelons also explain the ground rule in this system, namely the transfer in another legion. Promotion in the same legion depended whether or not the equivalent post in the next cohort was vacant, thus transfer was more useful and more encountered. The possibility of such a promotion is detailed in CIL VI 3584 by *promoveri in legion eadem*. This system lost its meaning when soldiers from the provinces are promoted and as the army becomes more and more barbarized. The naming of centurions in the 3rd century as *ordinarii* and *ordinati*, similar to centurions from the *numeri*, expresses this alteration of the system.

[534] Domaszewski 1908, 95-97.

[535] Allen 1908, especially 4-5, 6-10.

[536] Radin 1915, 300-310.

[537] Parker 1926, 45, 51-52; Parker 1928, 277-283.

[538] Le Roux 1982, 292; Keppie 1984, 151; Strobel 1987, 204-208. Strobel 1988, 43-45. This procedure is opposed by Richier 2004, 488.

[539] Wegeleben 1913, 23.

[540] Richier 2004, 464.

[541] Wegeleben 1913, 24-25.

centurions *ex caliga* and those *ex equite Romano*, the latter needing such a career in order to become *primus pilus* and advance further. It is of course true that a centurion *ex caliga* can advance quickly, proving his skill and training for a superior career but careers as centurions in cohorts II-X went forward without any clear rules, only skills and merit being decisive for the length that they occupied these posts and for promotion to higher positions of centurion.[542]

As far as the terms *succedere, succesione promoveri* and *promoveri*, they only establish the manner in which the centurion advances.[543] The terms *hastatus, princeps* and *pilus* are not ranks but attributes, that their position can be clearly established in the legion. In some cases the last position to be held is mentioned, firstly because it is fresher in the memory and because it was obtained after a long period of service and in the end this is what matters. The importance of the post *pilus prior* in the cohorts II-X, as a virtual commander of the cohort, is not real because it is the least mentioned post in inscriptions, while it should be just as important as that of *primus pilus*. The superiority of *priores* over *posteriores* also cannot be assumed.[544]

Based on what Wegebelen has proven so far at least two conclusions can be drawn. 1. The general name of *centurio* refers to any post from the cohorts II-X. 2. Transfer between legions does not mean promotion because in all cases they remain in the cohorts II-X. Plus, the identity in unit strength between the cohorts II-X cannot allow for a difference between several commanders each leading the same number of soldiers.[545]

He then focuses on the problematic of the centurions of the first cohort, in his opinion the only superior rank of centurion. He assumes as real its double number of soldiers but opts for a number of six centurions.[546] Concerning their internal hierarchy he claims that the centurions *posteriores* of the first cohort must be equal in rank because there is no mention of a *primus pilus posterior* in any inscription. Thus, the second stage in the career of centurions started with access to the first cohort on one of the posts of *posteriores*, from where, based on their performance, they could become *primus pilus*.[547] Vegetius says that only the first cohort is *milliaria*, not that there would be any difference between its *centuriae*. It can be presumed that the three *centuriae* of *posteriores* have 100 people each, just like the other normal ones from cohorts II-X. In this case *I hastatus prior* should be superior to the three *posteriores* although it also seems that he too commands a *centuria* 100 men strong. In a series of inscriptions[548] this post appears separate, simply written as *hastatus*, much more often than the posts of *posteriores* from the first cohort: this might indicate a superior rank, as confirmed by Caesar.[549]

All this evidence can show us the promotion principle for the first cohort. We must differentiate clearly between the two classes of centurions from the first cohort, the lowest ranking of the *priores* obviously being *hastatus*.[550] The posts of *posteriores* were probably easy to reach while the other high-ranking three were reached following serious efforts and exceptional performance. The immediate superior rank is *princeps*,[551] the last one before the position of *primus pilus*. Thus, there are five ranks of centurions: those from cohorts II-X, the three *posteriores* from the first cohort and finally the *hastatus*, the *princeps* and the *primus pilus*. It was mandatory to hold a position as legionary or praetorian centurion before *primus pilus* but not before *princeps*.[552] Any worthy centurion from the cohorts II-X could be directly promoted to *primus pilus*, without holding posts as *hastatus* or *princeps*. The difference in the number of troops these three centurions commanded, expressed by Vegetius[553] is in fact reflected in the difference in rank. The reason why these three posts are so different from other centurion posts from cohorts II-X, with their corresponding names, is a simple one and resides in the fact that in the case of those three from the first cohort the word *centurio* is not mentioned, only the post. We might be inclined to believe that these three do not even belong to the body of centurions, because of this omission, but we can state that the difference in rank between them and the *posteriores* from the first cohort is much greater than that between the latter and the centurions from cohorts II-X.[554] As far as the notion of *primi ordines* is concerned, he associates it without any doubts to the centurions from the first cohort.[555]

The one who followed closely, cultivated and at the same time developed this theory, while passing it on to his students, is E. Birley. Following his comprehensive study regarding the origins of legionary centurions he moves on towards studying hierarchy and promotion systems. To this aspect Birley prefers Wegeleben's theory: that all centurions from cohorts II-X are equal in rank, only the *primi ordines* being superior.[556]

B. Dobson, who calls himself in the preface of his book concerning the *primi pilares*, 'a student of Birley', successfully pushes forward in the direction of study he opened for the research of centurionate. First of all, in the critic edition of *Rangordnung* he states that the hierarchy of centurions is still open for discussion. Still, he agrees

[542] Wegeleben 1913, 26.
[543] Wegeleben 1913, 32.
[544] Wegeleben 1913, 32-34.
[545] Wegeleben 1913, 35.
[546] Wegeleben 1913, 36-40.
[547] Wegeleben 1913, 42.
[548] CIL II 1681, 4146, 4147, III 2836, 3846, 14477, VIII 2825, ILS 484.
[549] *BC* 1.46: a *hastatus* reaches this position from a lower one, due to his bravery. Wegeleben 1913, 43-44.
[550] CIL III 8112.
[551] CIL IX 4122 and XI 5273. Wegeleben 1913, 45.
[552] Wegeleben 1913, 47-48. Vegetius' note, *Epitoma* 2.8, where he speaks about promotion from *princeps* to *primus pilus*, must not be seen as a rule but as an option, especially because *primipili* who mention a previous post did not hold the post of *princeps*.
[553] *Epitoma* 2.8: *hastatus* – 100, *princeps* – 200 and *primus pilus* – 400.
[554] Wegeleben 1913, 49-50.
[555] Wegeleben 1913, 51-54.
[556] Birley 1964, 206-215.

with Wegebelen who contradicts all opinions expressed by his predecessors and claims that Domaszewski's theory has weaknesses.[557]

In any case this second variant of centurion hierarchy is accepted by a large number of academics interested in Roman military studies that have taken and inserted it in their studies.[558] In this situation that seems unsolvable, part of the academics were content with simply accepting that the hierarchy of centurions is yet to be discovered and focused themselves on other particular aspects of the centurionate.[559] More recently others tried to elucidate the meaning of some of the most complicated accounts given by ancient military historians.[560] All these examples only serve to confirm that becoming a centurion was regarded as a real promotion as compared to the numerous transfers that might be encountered in a centurion's career.

Unfortunately, with regards to the hierarchy of legionary centurions we still cannot draw a final conclusion. It is probable that similar to the *principales*, where we encounter a multitude of functions and only three ranks, the situation is the same for the centurion: a multitude of posts that are differentiated in between not by salary but by attributions. The epigraphy of the Principate only complicates this matter by disguising several titles under the general formula *centurio legionis*. Now even those who become *primi pili* clearly explain their career path, being able to occupy between one and nine previous postings as centurion. In fact, besides the other two centurions *priores* from the first cohort – *hastatus* and *princeps*, the *primi pili* give the same name to all their previous centurionates. Add to this the wars that must have favored transfer and promotion.[561]

As far as the *primi ordines* are concerned their identification with the centurions of the first cohort is confirmed by ancient sources[562] that place them hierarchically in between tribunes and centurions.[563] Caesar[564] clearly refers to the *primi ordines* as being the centurions of the first cohort, Tacitus,[565] when describing Vitellius' troops entering Rome mentions the *primi ordines* in front, with the *praefectus castrorum* and the *aquila* and Arrianus[566] groups around the eagle the legate, the tribunes, the *praefectus castrorum* and the five centurions who lead the first cohort. Oddly enough they are the only ones to still serve as officers in the legions from the time of the Dominate.[567]

II.1.3. Excurse. The centurionate in Rome

Centurions of the *vigiles*. A number of 190 centurions from *vigiles* cohorts are known from epigraphy but for many of them we know nothing but the name. Most are dated in the 3rd century because they are mentioned on the *latercula cohortis V* from *mons Caelius* and *grafitti* from the *excubitorium* in *regio XIV*.[568]

Starting with the 1st century and until the reign of Severus *vigiles* centurions come from the praetorians recalled into service. In the 3rd century we can notice a relative flattening of legionary and urban careers, one being able to reach the *vigiles* centurionate from a multitude of posts varying in rank.[569] There are two options for further promotion: towards the other centurionates in Rome, then *primus pilus* and eventually an equestrian career or the end of the career. Still, there are many who advance no further than this second post.[570] Length of service for a *vigiles* centurion depends in principle to the number of vacant posts; more precisely half their number and many of them were occupied by *ex equites Romani*, directly reaching the posts after service in the legions.[571]

There is little information concerning the functions of *vigiles* centurions but by their very nature they are more closely connected with the soldiers than the tribune. Taking into account their past service in the praetorian cohorts or in legions it stands to reason that most likely they were not specialists in fire control but rather officers who had between the 2-4 years of service to learn the job. The centurion gives the name of his *centuria*, not defined in any other way and some centurions also command the vexillation from Ostia. The relation between *vigiles* centurions is close,[572] as discernible also from the group dedications they make.[573]

Centurions of the *urbaniciani*. The officers of the *cohortes urbanae* in Rome are the centurions and the tribunes, the latter occupying the second place after those from the praetorian cohorts. Following analogies found in inscriptions,[574] where urban tribunes are written after the praetorian ones it has been concluded that this must also be the relation between the centurions from the two types of troops.[575] Many centurions from the *urbanae* cohorts are in a similar situation with the tribunes of these units.[576] The centurions from the *urbanae* cohorts are identical in rank. One can reach this centurionate in two ways: as a former praetorian soldier recalled for service, who becomes centurion in the *vigiles* and then the *urbani* troops or as a legionary centurion *ex equite Romano* who comes to Rome directly on a post of urban centurion and has better perspectives for promotion. Most urban

557 Domaszewski-Dobson, 1967, xxiii; Dobson 1974, 393-408.
558 Watson 1969, 88, 119; Junkelmann 1986, 111; Le Bohec 1994, 44, Campbell 1994, 46.
559 For example M.P. Speidel.
560 Richier 2004, 489-496.
561 Richier 2004, 491.
562 Vegetius, *Epitoma* 2.7, 2.8; Tacitus, *Hist.* 3.22.8, 3.49.3, *Ann.* 1.32.3; Hyginus, *De munit. castr.* 1, 3; Amm. Marcellinus 16.12.20.
563 Frontinus, *Stratag.* 20.4, Caesar, *BG* 6.40.7, *BC* 1.46.4.
564 *BG* 1.41.3, 5.28.3, 5.44.1, *BC* 1.77.2, 3.53.5.
565 *Hist.* 2.89.
566 *Ektaxis* 1-10.
567 Richier 2004, 492-494.
568 Sablayrolles 1996, 162.
569 Sablayrolles 1996, 163-164.
570 Sablayrolles 1996, 165-166.
571 Between two and eight years. Sablayrolles 1996, 167.
572 CIL VI 2961 – *sibi collegisque suis et futuris.*
573 Sablayrolles 1996, 173-174.
574 CIL VI 1009.
575 Freis 1962, 77.
576 Mench Jr. 1968, 501.

centurions are former praetorians (thus falling into the first category). Promotion is always towards the praetorian centurionate,[577] suggesting they are only passing through this post on their way to a higher centurionate. Those who promote from the military ranks have a slower career, probably ending with a centurionate or *evocatio*.[578]

Advancing quickly through the three centurionates in Rome depended on the availability of posts for each of them. There are 49 *vigiles* centurionates and only 24 *urbanae*, because the units from Lugdunum and Carthage had different paths of promotion. The praetorians have 60 centurionates starting with Domitian and they are occupied with soldiers from two sources: a part of them were transferred legionary centurions[579] and the rest had very long careers without ever leaving the praetorian cohorts.[580]

It can be noticed by analyzing all their careers that service length for the urban centurion varied. Except for those with better possibilities, who promote very quickly, there are others who remain on that post for a longer time. However, Durry's statement, that passing quickly through the urban centurionate is the rule, is acceptable.[581]

Praetorian centurions. Although their hierarchy is not as complex as that of legionary centurions it still poses problems that have not been solved so far. Unlike legionary centurions, who can come from a multitude of sources, praetorian ones are praetorian soldiers, promoted, then recalled for service, or men who have served as legionary centurions and following that have passed through the *vigiles*, *statores* and *urbaniciani* centurionates. Another pool of recruitment is represented by the knights, presumably excluded from the equestrian order[582] for military service and who are directly appointed centurions. The area of origin of praetorian centurions is the same as that of all praetorians, meaning Italy and the most Romanized provinces, up until Severus.[583] In the 1st century all praetorians are Italians but starting with the 2nd century soldiers from Romanized provinces, such as Macedonia, are allowed entrance, everything culminating with the cosmopolitism from the reign of Severus.[584]

The prospective career of praetorian centurions was more illustrious than that of their legionary counterparts. They joined the legions on a post of *princeps*, then became *primus pilus* and could have access to equestrian functions even as young men, unlike their legionary comrades who were old upon reaching this stage. The limits of the praetorian centurion were those he himself set.[585]

By comparison to legionary centurions, with their names and position written in their entitling the praetorian ones have no such indication. Besides, the *centuriae* do not have a common name, nor are there any *primi ordines* in the cohorts from Rome and thus no *primus pilus* either. Thus, in the praetorian cohorts all centurions are equal in rank and this plus the lack of the eagle is meant to differentiate the *praetorium* from the legion.[586] Although they have no internal hierarchy, they have special posts meaning they are assigned away from their unit on different tasks: *exercitator equitum praetorianorum, exercitator equitum speculatorum,* as well as two posts that are somewhat unclear: *princeps castrorum* and *trecenarius*.[587]

All other praetorian centurions are equal, with no different ranks, no name, thus we cannot differentiate them; this means that none of them served as praetorian centurion more than once, after that being free to pursue his own advancement path. This aspect probably has historical reasons. In the legions the names of the three battle lines remain because of the administrative need to create a hierarchy, the differences in weaponry disappearing after the reform of Marius, but praetorian cohorts initially had six, then ten centurions, who had no need for such separations. They rank higher than their legionary counterparts but lower than the *primi ordines.* Even after their number is increased praetorian cohorts are still independent of each other, thus there was to need to import the structure of the legion. Probably it was decided against excessively favoring and empowering a praetorian centurion so that hierarchy or insubordination problems would not arise.[588] The cohorts have a single standard, each one bearing the *imago Caesaris* and have no eagle in order to completely distinguish them from the legion.[589]

Still, there are three more other classes of centurions 'without rank': the *exercitator equitum praetorianorum, exercitator equitum speculatorum* and *princeps praetorii/trecenarius*, posing a lot of problems.[590] Domaszewski tries to classify and to differentiate between them but to no great success, being followed by Durry. He proposes the *centurio speculator*, the commander of the body of *speculators* from the

[577] Freis 1962, 77.
[578] Mench Jr. 1968, 501.
[579] Freis 1962, 81.
[580] A fast-paced career is that of M. Vettius Valens (CIL XI 395), who in 23 years moves from *beneficiarius praefecti praetorio* to *procurator Lusitaniae*. Freis 1962, 79.
[581] Durry 1968, 135. For the urban cohorts from Carthage and Lugdunum see Duval, Lancel, Le Bohec 1984, 57; Freis 1962, 87; Bérard 1995, 373-382.
[582] Cf. Domaszewski 1908, 80 and Richier 2004, 463.
[583] Durry 1968, 133; Passerini 1939, 78-79.
[584] Durry 1968, 134.
[585] Passerini 1939 79-80.
[586] Durry 1968, 136.
[587] Passerini 1939, 82.
[588] Le Roux 1992, for AE 1990 752-755, Fiésole, proposes the wording *centuria Vernini prioris*, of the *VII praetorian* cohort, perhaps indicating a certain hierarchy between the *centuriae* from a praetorian cohort at least from the beginning of the 3rd century as there is no prior evidence for this. Multiple clues about the dating of the inscription in the 3rd century, as well as the recruitment of Germanics in the *praetorium*, confirm Severus' changes in matters of military policy. Le Roux 1992, 261-262.
[589] Passerini 1939, 83-88.
[590] Durry 1968, 137.

praetorian cohorts starting with the reign of Nero. Following that he takes the title *trecenarius-CCC*. It ranks higher than the praetorian centurion because the holder commands this body of 300 elite knights who are part of a mounted guard.[591] Before the post of *trecenarius* a praetorian centurionate is always observed and the two posts must not be combined into a single one.[592] The post must have been a very honorable one because all *primipilares* who have held it always mention it. Praetorian centurions are also known to promote towards *princeps castrorum/princeps praetorii Augusti* or *exercitator equitum speculatorum*. The post of *princeps praetorii* refers in this case to the commander of the *officium imperatoris*, mentioned in Hyginus' description of the camp, made up of *evocati*. Starting with Trajan and following the creation of *castra peregrina* and the creation of the *princeps castrorum peregrinorum*, the title is changed to *princeps castrorum (praetorianorum)*.[593] He also proposes a kind of praetorian *primi ordines*, defined rather vaguely.[594]

After Domaszewski, the position of the *trecenarius* has been debated upon and it is no longer certain that the praetorian centurionate precedes it. Promotions from *trecenarius* to a legionary centurionate and the other way around are known. *Princeps praetorii* and *princeps castrorum* outranks the *princeps peregrinorum*, who appears from the time of Trajan. The exact relations between the centurionates in Rome and the *primi ordines* cannot be proven.[595]

Coming back to the centurion promotion scheme from Rome Domaszewski offers a pretty straightforward direction: *centurio coh. vigilum – centurio statorum – centurio coh. urbanae – centurio coh. praetoriae*.[596] In any case the soldiers from the cohorts in Rome could hold three centurionates, one for each kind of troop, in ascending order. This path was allowed also to those who are legionary centurions and come in Rome without serving as *vigiles* centurion, this post being equivalent to their legionary centurionate. This analogy might be a clue for understanding the significance of the *trecenarius*.[597] Finally, there had to be a hierarchy in the ranks of the praetorians but perhaps resembling more that of the *centurio princeps* from the auxiliary cohorts.[598]

Centurions of *castra peregrina*. Besides informers – *delatores*, the emperors from the 1st century made very good use of the praetorians, especially the tribunes and the centurions for missions of intelligence, arrests and executions for treason. From the beginning, any organization of the Roman secret service included the body of centurions. The most efficient soldiers for this kind of missions, the *frumentarii*, while detached to Rome were garrisoned in the *castra peregrina*, under the command of a superior centurion, *princeps peregrinorum*, who answered to the prefect of the *praetorium*. Most *frumentarii* managed to advance in ranks in this manner, towards the peregrine centurionate. From here it was easier to accede to legionary centurionate or the other centurionates in Rome, then obviously to *primus pilus* and perhaps even beyond that.[599]

A number of soldiers and officers stay in the *castra peregrinorum* apparently reunited under the name of *peregrini*: *frumentarii*, *speculatores* and other ranks from the *officium consularis* detached to Rome. Their officers are, in ascending order: *subprinceps peregrinorum*, *princeps peregrinorum*, besides them we find mentioned the *centuriones deputati* and *centuriones supernumerarii*.[600] Even the highest rank, the commander of the *castra*, was on the same rank position as a provincial legionary *hastatus*. In the 3rd century all of them have equal ranks with the praetorian centurions and have the same access to the legionary centurionate and the position of *primus pilus*. In the 2nd century these posts are not mentioned in the career of the *primus pilus*, thus they did not matter in the upper hierarchy of the army, but in the 3rd century such executors had great success with the emperors, many of them obtaining high honors following assassination missions at the orders of the Eastern despots. Indeed *castra peregrina* and its mechanism can be dated only post-Trajan, but the centurionates here are rather a delegation that a separate rank.[601] At any rate the centurions serving on the posts of *princeps peregrinorum* and *subprinceps peregrinorum* are considered to have a high rank; in fact they command all soldiers from the peregrine fort.[602]

In reality, in order to make military service and military careers more attractive Severus had to modify the promotion rules for soldiers and centurions thus, contrary to the established situation from the 2nd century, many soldiers benefit from a great career, ending with procuratorships or even the post of *praefectus praetorio*. Thus there was a democratization of the military career,

[591] Domaszewski 1908, 99.
[592] AE 1906, 73 – between the two a *princeps castrorum* is inserted.
[593] Domaszewski 1908, 101; Durry 1968, 137-138.
[594] Domaszewski 1908, 102. Dobson, Breeze 1969, 119 considers the *trecenarius* a kind of praetorian *primus ordo*. On the other hand, Passerini says that the *trecenarius* is an obscure rank that cannot be attributed to any cohort or position. It can be considered a special rank of praetorian centurion, the highest ranking one, this expression replacing any sign of previous career in the urban centurionate and the literal meaning of the word would indicate the commander of 300 soldiers. The problem appears when these *trecenarii* are joined with a legion. The hypothesis stating that he is the commander of the body of *speculators praetoriani*, thus an evolution of the term *centurio speculatorum*, has no evidence supporting it and we must keep in mind that the two terms are used in the same time period. Even if we do not know exactly what it means it existed both in the praetorian cohorts and in the legions and we can only say that it is inferior to the *princeps legionis* and thus to the *primus pilus*. Passerini 1939, 90-94; Durry 1968, 139.
[595] Domaszewski-Dobson 1967, xxvi; Mann 1983a, 138-140; Richier 2004, 453-456; Christol-Demougin 1986, 198; Petolescu 1995, 245-247.
[596] Domaszewski 1908, 103.
[597] Mann 1983a, 136-137, equivalent to *a militiis*?.
[598] Passerini 1939, 95.
[599] Sinningen 1961, 66-67, Rankov 1990, 178.
[600] Baillie Reynolds 1923, especially 171-179.
[601] Domaszewski-Dobson 1967, xxvii-xxviii.
[602] Mann 1988, 149.

the limits being now imposed by one's own talent and possibilities rather than social origin.[603]

Centurions of the praetorian fleets. Taking into account all examples of petty officers who are mentioned it can be stated that fleets are organized from this point of view just like land-based armies, with the exception of special technicians.[604] Still, Domaszewski's supposed theory of total equivalency is much too general. He argues that the centurions from the praetorian fleets, another neglected category during the first two centuries of the Principate, end up having many more opportunities in the 3rd century. The equivalency between *navarchi* and *trierarchi* with legionary centurions starts from Antoninus Pius-Marcus Aurelius[605] and in the 3rd century many of them have careers in the legionary centurionate. The *navarchus princeps* is the equivalent of the *princeps legionis* and was directly promoted to the post of *primus pilus*,[606] and Marcus Aurelius and Lucius Verus put the *trierarchi* on an equal footing with *tertius ordo* from the legions.[607] Unfortunately the situation of hierarchies in the fleets is not nearly as clear, sometimes being impossible to solve.[608]

[603] Smith 1972, 494.
[604] Reddé 1995, 151.
[605] CIL X 3340.
[606] CIL X 3348, though it refers to *primipilus legionis I Adiutricis*.
[607] Domaszewski 1908, 106.
[608] Domaszewski-Dobson 1967, xxviii. Sander 1957, 353-356, 362-364; Reddé 1995, 152; Saddington 2007, 210-211; Oorthuijs 2007, 170-180.

II.1.4. Special functions of the centurions

Besides his primary military role the centurion proved to be such a well-trained and loyal officer as to be used for an impressive variety of tasks. We find included administrative and economic tasks, maintaining internal security and public order, circulation, intelligence and even diplomatic duties. A somewhat complementary task to his role as an officer was that of commanding and administering different military units or contingents of soldiers, created on the spot or sometimes definitely, for certain purposes.[609]

In the provincial armies they are used as *praefecti civitatium*, especially after the harsh lesson of the rebellion from Pannonia when Augustus put the whole of Illyricum under a very strict military administration system led by legionary centurions and knights (*praefectus civitatium Colapianorum* was a *princeps legionis*). Centurions are especially used to replace knights, but at the beginning of the Principate there are no clear rules established.[610] The post of *praepositus numeri* seems to be held for a longer period of time, the centurions becoming the traditional commanders of these barbarians. In the 3rd century, when such differences between the units disappear, the post transforms into a *praefectus* one and then generalizes as a *praepositus*. In bodies of soldiers taken away from legions or other troops in order to be sent on missions, centurions occupy the command post as *praepositi vexillationis*. They are also tasked with local policing, public order or supervising different infrastructure projects and in the administration they can even be *iudices*. They are used in administering the *territorium legionis* or as inspectors for the imperial roads, for traffic control. The tasks not connected to provincial armies can only be assigned directly by the emperor[611] and are connected with major infrastructure projects or imperial enterprises.[612]

As to the aspect of extraordinary command tasks, R. Saxer analyzes in detail the commanding officers in his study about the vexillations of the Roman army. Firstly he divides these vexillations according to their role, meaning warfare or building. Centurions seem to be excluded from the command of vexillations that go to war, especially because of the number of troops: they needed officers of equestrian or senatorial rank.[613]

[609] Richier 2004, 539.
[610] Domaszewski 1908, 107.
[611] Plinius, *Ep.* 77, 78
[612] Domaszewski 1908, 108-109.
[613] Up until the Marcomannic Wars legionary vexillations could only be led by officers of senatorial rank, meaning tribunes, legates or governors. Tribunes and legates could lead them only if detachments were taken from legions stationed in the province where they serve. Before Severus there are only two cases of *primipilares* commanding legionary vexillations, but starting from his reign they can also officially lead them. From the time of the Marcomannic Wars more and more knights are admitted in the senatorial order and will obtain extraordinary commands of vexillation. From Gallienus onwards only officers of equestrian rank can still command them and they have different titles: *protector*, *praepositus*, *dux*. Auxiliary vexillations are led by officers of equestrian rank: *praefectus/tribunus cohortis*, *tribunus legionis angusticlavius*, *praefectus alae*.[613] At the beginning of the Principate the

Legionary building vexillations are almost always headed by centurions. In theory the number of troops commanded by a centurion must be no more than the number of soldiers in a *centuria* – 80 men. Centurions commanding building vexillations are entitled in principle *praepositus* but there is also the case when the *sub cura* is added to the title *centurio*. It was believed that there is a difference between the two terms and that the latter would refer in fact not to effective command but to supervising the construction activity. In fact *sub cura* has a much larger meaning, including a supreme command over a detachment or office and the simplest meaning is supervising and organizing the building of a monument. In these situations the command of these centurions is certain, being combined with the supervising function.[614]

As to the effective command of auxiliary troops at the beginning of the Principate it has been proven that, first of all, it cannot be associated to any functions of equestrian rank,[615] and, second, that such posts were almost exclusively occupied by *primipilares*.[616] Still, a type of auxiliary troop has been found which, during the period of the Principate, a centurion could command. As a rule, officers commanding *numeri* were legionary centurions, entitled *praepositi*. There are also cases of auxiliary officers but only as exceptions. The term *praepositus* was not reserved solely for *numeri* but it appeared in every command of auxiliary troop given to a legionary centurion, probably because of a lack in equestrian officers at a certain moment; they are replaced by *praefecti* and *tribuni* of equestrian rank from the end of the 2nd century onwards.[617]

Coming back to provincial armies, but this time focusing of the governor's staff, centurions were given a lot of tasks here, out of them several tasks of command stand out as pretty prestigious and profitable. As a rule legionary centurions with the title of *praepositus* or *curam agens* command the governors' *pedites* or *equites singulares*. Both titles are characteristic of temporary commands and are fit for the status of the governor's guard, as they are detachments rather than complete troops. In provinces without legions even a *decurio alae* or a *centurio cohortis* could lead them, receiving the title *princeps praetorii*.[618] Besides them, legionary centurions were also appointed to supervise the training of the *equites singulares* as *exercitatores*. In one case the two functions, of commander and instructor, seem to be combined (*praepositus simul et campidoctor*). Probably more than one instructor was attached to the *singulares* from a province, following the analogy of the *equites singulares Augusti* who for 1000 men had four centurions *exercitatores*. We do not know the rank of these centurions, some could be *primi ordines*, others only *evocati*.[619] As to the role played by centurions in the imperial category of the *equites singulares* it can be noticed that Trajan abandoned the anonymous Greek instructor who served in Domitian's mounted guard and returned to the custom of bringing legionary centurions with a lot of experience and a high ranks (*exercitatores*) for the training of his *equites singulares* and, through them, of all Roman cavalry units.[620]

There are many changes in the Roman army in the 3rd century, most of them being attributed to Gallienus, although they begin with the reign of Severus. Still, the title *protector* is encountered for the first time under Gallienus. Epigraphic sources of the period reveal to us an increasing number of *protectores*,[621] leading to new conclusions about this body of soldiers. It can be considered a group of soldiers attached to the emperor's guard, probably replacing the *equites singulares Augusti*.[622] This title was granted to centurions who served in the units of the mobile army created by Gallienus to counteract the numerous breaches the barbarians made on the European *limes*.[623] Thus the title of *protector* can be associated to the earlier one of *comes* that was granted to officers serving away on campaigns alongside the emperor. It is not reserved only for centurions and it can be granted to all officers who are part of the mobile army.[624]

In provinces, centurions have a multitude of tasks, of military, paramilitary, economic and juridical nature. Some of them work in the governor's office. Especially in the 3rd century we encounter epigraphic evidence concerning centurions who head provincial *officia* – *princeps praetorii*, probable aided by an adjunct – *optio praetorii* and an assistant – *adiutor principis praetorii*.[625] In these *officia* centurions could also be used for diplomatic missions, as *negotiatores*[626] and those specialized in translating barbaric languages of local dialects were used as *interpretes*, in barbarian centres.[627] In Germania, where each province had two legions each, the geographic proximity of one legion's fortification to the governor's headquarters was not a decisive factor in recruiting necessary personnel.[628]

---

title *tribunus* or *praefectus vexillationis* is also encountered both for knights and *primipilares*. Starting with the Marcomannic Wars only the titles of *praepositus* and *dux* will predominate. Saxer 1967, 122-123.

614 Saxer 1967, 129-131. Smith 1979, 263-278. *Decurio alae* played a very important role in provinces without legions because in the absence of a legionary centurion he could also serve as *praepositus*.

615 Domaszewski-Dobson 1967, xxviii.

616 Birley 1978, especially 262-269.

617 Southern 1989, 98.

618 Speidel 1978, 26-27.

619 Speidel 1978, 28-29. But as we can see from the career of M. Calventius Viator, it can be a very profitable function because of the close contact with very important persons. Speidel 1978, 29-30.

620 Speidel 1994, 42, 64-66, 69. Speidel believes Viator (CIL III 1904=IDR III/2 205) foiled Nigrinus' plot against Hadrian. (Dio 69.2.5, *HA*, *Hadr.* 7). Speidel 1978, 28-29 and Speidel 1994, 47-48. *Contra*, Petolescu 2005. Other approach in Cupcea 2013 and Cupcea 2015, forthcoming.

621 CIL XI 1836, III 3228, AE 1920, 108, CIL III 3529, 3424, AE 1965, 9, 114.

622 Christol 1977, 393-395. For further explanations see the career of Traianus Mucianus: IGBR III/2 1570, 1568 and 1569, Thracia. Christol 1977, 397-404

623 AE 1954, 135.

624 Christol 1977, 405-407.

625 Austin, Rankov 1995, 150.

626 Tacitus, *Ann.* 13.54-55, 4.72.3, Dio 71.15-16, 72.2.4

627 CIL III 10505, 14349^5. Austin, Rankov 1995, 26-28.

628 Richier 2004, 548.

Several academics that have studied the economic history of the Roman Empire or its administration have discovered the intensive use of soldiers, including centurions for these tasks. From the middle of the past century, H. Zwicky considered that in the case of centurions the military aspect of their rank took precedence thus they were used more often as *praepositus vexillationis*, charged with guard duty, policing and administrative tasks *etc.* Many times however they command building detachments, especially in the province where their army is camped[629] or mines and quarries.[630]

In the area of the Danube the best-known mining area is Montana, where we find vexillations from the local legions, after that to be replaced by a *numerus civium Romanorum* under the command of centurions from those same legions.[631] In spite of the large mining infrastructure from Dacia epigraphic signs of the military presence are few. Except for a few *beneficiarii consularis/procuratoris* and *librarii* at Ampelum, perhaps proof of a contingent stationed here, it is likely that the security of the area was ensured by the proximity of the forts from Apulum and Micia.[632] Still, the building activity of the legions *XIII Gemina* and *IIII Flavia* can be observed at Ampelum on tile stamps.[633] This is true more so because lately several stamps of *XIII Gemina* as well as the remains of an official building have been found at Alburnus Maior.

In the Eastern provinces, the centurions from *XVI Flavia Firma* supervise road reparations[634] besides ensuring the security of the roads or of the defense or surveillance towers.[635] All epigraphic monuments mentioning legionary centurions in the field suggest they were tasked with ensuring maintenance and security of the transport ways in southern Syria. Also in many cases they are mentioned as commanding garrisons, some of them smaller, sent in cities around Judaea to fight charlatans and false prophets or as *regionarii* centurions in a district to ensure security.[636] Also it has been proven that starting with the 3rd century military personnel played a very important role in gathering taxes.[637]

There is a great number of evidence that they were used as *iudices*[638] to settle property limits *etc.*, in which case they are agents of the governor.[639] The most encountered judgment that a centurion had to make was the appeal to the emperor, the right of every citizen condemned to death.[640] Centurions are interested in crimes that might endanger the social-political security, in organized crime or in the corruption of the fiscal system. They do not seem very interested in private matters, as long as they do not lead to violence. Still, they are approached with all kinds of problems. Usually they are asked to bring the criminal to justice or ensure physical protection.[641]

All throughout the Empire perhaps most of them are used in policing and ensuring safety missions, especially in the wide-ranged system of military *stationes* along roads, important objectives *etc.*, for ensuring the safety of traffic and commerce. Such mentions are found very often, at least starting with the reign of Commodus. We do not know with certainty the organization of these *stationes*, most mentions referring only to the commander or their personnel, most frequently of consular beneficiaries. It is likely that a centurion heads the personnel or that a centurion is sent on mission for the more serious tasks.[642] It depends on the size and importance of these *stationes*; the smaller one can be left in charge of *immunes*. Those who serve here somewhat continuously are known as *stationarii*. Although these men seem to be dispatched by the governor, the safety of the roads, of goods and of information is still ensured by the central administration,[643] even more so because some *stationes* had to collect custom taxes.[644]

In some of these instances, they can bear the title *regionarius*, which means that they are in charge of a territory organized as a *regio*, placed under military command. In this case the centurions are in fact commanders of a military squadron of *stationarii* or other soldiers, ensuring security and the reign of the law.[645]

In diplomatic missions legionary centurions or *primipili* are used several times. Tiberius sent a centurion to the Thracian princes Rhaskuporis and Kotys; centurions were also used in the Eastern wars of the Julio-Claudian dynasty, during the Civil War of AD 69 and in the Flavian period.[646] Sometimes legionary centurions are even sent beyond the *limes*. An *interprex legionis XV idem centurio negotiator*,[647] is buried in Slovakia, north of the Danube where he was probably sent at the *Quadi*.[648] An *interprex* Alexander, who is mentioned in the East around AD 169-171, is sent on an official mission by Avidius Cassius along with some legionaries and one centurion. When negotiating peace with the *Marcomanni*, Marcus Aurelius imposed that every tribe gathering should take place only once a month, in the

[629] Zwicky 1944, 76; Hirschfeld 1963, 244; Hirt 2010, 22, 169, 293-294.
[630] Hirt 2010, 170-172.
[631] Speidel 1984, 185-187; Hirt 2010, 70-71, 193.
[632] Hirt 2010, 195.
[633] Piso 2000, 208-213.
[634] CIL III 199-201.
[635] IGR III 1113, 1114, 1121. Isaac 1990, 135; Pollard 2000, 96.
[636] Davies 1989, 56; Isaac 1990, 135-136.
[637] Sperber 1969, 186-188. In opposition, the centurion mentioned in Mathew's Gospel 8:5-13 is not considered a Roman officer but rather one belonging to the local army. Saddington 2006, 141-142.
[638] Pollard 2000, 91.
[639] CIL III 9832.
[640] Zwicky 1944, 84. Once most inhabitants of the Empire are granted citizenship by Caracalla, they increase considerably in number. Richier 2004, 552.
[641] Alston 1995, 91-92.
[642] Plinius, *Ep.* 77.
[643] CIL III 14349[3], from Aquincum, tells us that a centurion had an established area of competence, *curam agens pontium viae*, thus he was probably responsible for teams of *stationarii* in his area.
[644] Zwicky 1944, 83-86.
[645] See most recently Fuhrmann 2012, especially 220-230.
[646] Tacitus, *Ann.* 2.65, 13.9, 15.5, *Hist.* 2.58. Zwicky 1944, 80-82.
[647] AE 1978, 635.
[648] Isaac 1990, 399.

same place and under the surveillance of a legionary centurion, probably assisted by an interpret.[649] Another task that fell into the soldiers' hands, implicitly centurions, seems to be the capture of wild animals for the *venationes* shows. Many epigraphic sources mention *immunes venatores* or *ursarii*.[650]

Another problem that concerns military hierarchy, especially centurions, proved to be the specialist officers. Two particular categories stand out by the association of their title with the centurionate: *medicus* and *architectus*. As far as military doctors are concerned the detailed discussion should be made with regards to the categories of *immunes* and *principales*, categories that include a number of medically trained soldiers. Still, besides them, a category of *medicii ordinarii* stands out. The first to analyze the problem was J.F. Gilliam who stated that the association of the *ordinarius/ordinatus* with *medicus* or *architectus* denotes a civilian who temporarily served in the army, perhaps with a contract.[651] The idea that *medicus centurio* does not have tactical command can be false (or not universally true) and this is an example of the mixture of the technical and the tactical classes, the present character becoming *medicus centurio*, to promote to *princeps*.[652]

In the same registry, O. Stoll investigates the other association of the name of a specialist officer with the rank of centurion, *ordinatus architectus*. After many analogies, *ordinatus* refers here to a specialized officer who is enlisted in the normal *centuriae* but who is tactically independent, somewhat similar to the *equites legionis* or the artillerists.[653] It can be noticed that these technicians had the option of a tactical career if they so desired. The unusual position of technicians, especially *architecti*,[654] in the army seems to indicate their placement as specialized officers with a rank of centurion.[655] Finally, Stoll believes that all problems relate to our strict concept of *Rangordnung* that does not cultivate the difference between function and rank. In the same way that there are several types of *beneficiari* a *medicus centurio*, who does not seem to have tactical command, cannot be excluded from military promotion. In fact, officers like *medicus centurio* or *ordinatus architectus* make the difference between technical-administrative hierarchy and the tactical one; what we do not know is where they became intersected.[656]

[649] Dio 72.2.4. Peretz 2006, 462-463.
[650] Epplett 2001, 214-219; Devijver 1990, 143-145.
[651] Gilliam 1940, 145-148. The problem was discussed in detail by R.W. Davies, in his complete study regarding military doctors. Davies 1969, 89-91; Davies 1970, 87; Baker 2004, 44,48.
[652] Stoll 2003, 350-351.
[653] Stoll 2001, 331.
[654] Evans 1994, 147-148. Most are Greek civilians, Plinius, *Ep.*, 10.
[655] Stoll 2001, 367.
[656] Stoll 2003, 353.

II.1.5. Centurions' pay and social status

The one best corroborating the literary evidence with the epigraphical and papyrological ones up to this point is M.A. Speidel. In his study regarding payment grades in the Roman army he concludes that a centurion must have been paid 15 times the salary of a legionary, a *primus ordo* 30 times and a *primus pilus* 60 times that amount.[657] Taking into account the principles assumed by his predecessors he concludes that at the beginning of the Principate a legionary centurion must have received 4,5 times the salary of a praetorian (1,5 that of an *evocatus*). Another principle Speidel assumes is the constant increase in the centurion's salary directly proportionate to that of the soldier's, during the first three centuries of the Empire, because the hierarchy of the army and the promotion systems did not change radically in this period. Next he presents as probable the payments of centurions in the rapport 15:1 with that of soldiers, in *sestertii* per year, increased proportionally during the Principate:[658]

A recent recalculation maintains these numbers but changes the rapport with the legionary's payment. The difference between soldiers and officers is huge. Under Augustus a centurion obtained 13500 *sestertii*/year, 18 times more than a *miles*, a *primus ordo* double that of a centurion and a *primus pilus* double that of a *primus ordo*, thus 72 times the salary of a legionary. This rapport remained constant under the Principate up until Maximinus Thrax, a *primus pilus* having now a salary of 432000 *sestertii*/year. The incredible high salary of centurions as compared to that of soldiers is not only a mark of their importance in the army but also a mark of the importance of their professionalism for the security of the emperor and the Empire.[659]

[657] Speidel MA 1992, 101.
[658] Speidel MA 1992, 102.
[659] Herz 2007, 308. For the decoration system, Maxfield 1981, 184-186. For the right to *albata decursio*, Bennett 2006, 98-100; Speidel MA 2007, 268-270.

Rank	Augustus	Domitian (AD 84)	Severus (AD 197)	Caracalla (AD 212)	Maximinus Thrax (AD 235)
miles legionis	**900**	**1,200**	2,400	3,600	**7,200**
eques legionis	1,050	1,400	2,800	4,200	**8,400**
centurio legionis	13,500	18,000	36,000	54,000	108,000
primus ordo	27,000	36,000	72,000	108,000	**216,000**
primus pilus	54,000	72,000	144,000	216,000	**432,000**

Table II.1. Legionary pay under the Principate (*sestertii*/year), after M.A. Speidel.

The *praemia* received at discharge only served to increase the possible businesses centurions might have started while still enlisted. Q. Atilius Primus,[660] who goes from *interprex* to *centurio* and *negotiator*, sent in mission at the *Quadi* owned a considerable enterprise, which he leaves for his four freedmen to inherit. Centurions and soldiers are also involved in economic dealings and as witnesses at contracts,[661] meaning they received a percentage.[662]

Social status of the centurion. The social role of the centurion varies a lot, as seen in the previous chapters; he is probably the officer who best defines the Roman imperial society. But what about his own social status? Is he associated with the knights, or this is false even in theory? What is certain is that the Roman military environment allows for the greatest social mobility and the centurionate is the only way to legally advance on the steps of the social pyramid.

In the dissolution period from the end of the Republic we also encounter abuses[663] but admittance into the equestrian order following a professional military career was restricted only to those soldiers who, after a long and arduous service had become superior officers; this was one of Augustus' inventions.[664] The connection between soldiers and knights was created by the *primipilaris* who, as a soldier has passed through all ranks of *principales*, then through the centurionate to finally be *primus pilus*. Centurions are part of the third category of officers, next to soldiers, but the position is also open for knights and they do not have to give up membership in the equestrian order.[665]

The most spectacular social ascensions can be observed during the reigns of the emperors from the 3rd century. The best example for social automatic advancement is Pertinax, the son of a freedman, who becomes emperor.

Also, Vespasian's grandfather was a centurion that fought in Pompey's army.[666] The acceptance of a *decurio alae* in the equestrian order during the time of Severus Alexander[667] only serves to confirm a policy initiated by Severus who allowed *principales* to join the ranks of knights, consequently Gallienus opening this door for all soldiers. It is likely that by now the golden ring was no longer a symbol of this social class and some officers are allowed to join the knights because of bravery in combat.[668]

The duty of the emperor was to maintain and ensure social mobility under the guise of rewards for service to the throne, in the case of the lower classes,[669] or advancement to the upper classes also under the guise of service to the throne. In order to have legionary commanders and governors, the emperor had to convince the senators' sons to join the army as tribunes and to leave Rome for a long time. Those who were convinced were but a minority, eligible personnel had to be supplemented with *adlectio* of equestrian officers. They were the result of a policy to bring the equestrian order in the emperor's service. All available posts were not enough for the high number of those desiring them but those who succeeded could became provincial governors or even high prefects in Rome; following that, they themselves or their offspring could be received in the senatorial order. Thus, such titles were very attractive and up until Severus there was no problem in getting the posts occupied.[670]

When in service, the advantages of centurions were limited to the military domain. The emperor was not opposed to relations of camaraderie, choosing to keep group unity even after discharge as proven by the settlement of all veterans in a new colony, including the centurions and the tribunes, in order to create a

[660] AE 1978, 635.
[661] IDR I TabCerD VIII.
[662] Verboven 2007, 18-19.
[663] Such as the appointment of one of Sulla's centurions, L. Fufidius, as *praetor*. Sallustius, *Catill.* 37.6. Stein 1927, 208.
[664] Stein 1927, 135; Zwicky 1944, 89.
[665] Stein 1927, 136-139. This was proven following the clarification of the term *ex equite Romano*, which would have exactly to this renunciation because it does not appear in the inscriptions of knights that held the equestrian militiae. Still, *ex* is an apposition that appears in all military inscriptions where the previous rank, from where he was promoted, is mentioned. Zwicky 1944, 91; Speidel MA 1993, *passim*, after the analogy with *miles ex cohorte* or other similar mentions.
[666] Stein 1927, 238, 301.
[667] CIL VIII 21039.
[668] CIL XI 5992, X 5064. Stein 1927, 140-141.
[669] Not for all classes. In the Flavian age there was a case of a slave who was promoted to centurionate but because he had hidden his identity Domitian sentenced him to death. (Dio 67.13.1). Wesch-Klein 1998, 159.
[670] Dobson 1970, 99.

harmonious environment, developed in the time of Augustus.[671] Thus centurions remain even after discharge officers, spokesmen and even venerable parental figures for the soldiers.[672]

The literary activity of centurions is relevant for their individual level of culture and, although they come from different classes, for the centurions' general level.[673] It was unacceptable for a centurion to be illiterate. Those promoted from the ranks of the soldiers had been so after 15-20 years of service, sometimes on posts of secretary or accountants and the password was transmitted from one centurion to another in writing.[674] The army desperately needed a number of educated men,[675] maybe even to the point of knowing several languages.[676]

As far as the religious role of the centurion is concerned it seems it was rather monopolized by the *pimus pilus*.[677] Out of more than 80 *ex-voto* monuments discovered as set up by centurions from the Rhine area, 25 are set up in the name of the soldiers under his command. It falls to the centurion to mark down the fulfillment of the mission that a vexillation was sent on, but his religious role is strongly diminished in the legion, where he is overtaken by the *primus pilus*, the one whose honor it is to celebrate the *aquila* in the name of the whole legion. Being the closest officers to the soldiers, the centurions share with them the religious needs and in more than 48 private dedications it is noticeable that they share the pantheon with the soldiers. No private cult tendencies can be noticed at centurions, most dedications being for anthropomorphic divinities well established in the Roman military milieu. According to the chronology of the monuments it can be noticed that their incidence grows in moments when pressure is exercised on the *limes*. From the point of view of hierarchy, it can be noticed that the higher the ranks the lower the number of religious manifestations, the majority of dedicators being petty officers [678] (*beneficiarii etc.*).

A special category of military cults is that of *genii*. There are mentions of *genii* for most military units - *exercitus*, *legio*, *ala*, *cohors*, *numerus*, *vexillatio*, *cohors praetoria*, *cohors vigilum*, *equites singulares Augusti* and their subdivisions – *centuria*, *turma*. We find no such mentions for the urban or legionary cohorts but we cannot completely deny their existence. The time frame for these mentions ranges from the 1st century in Vindonissa[679] up until Diocletian, in Moesia Superior.[680] There are private or official dedications but we cannot speak about any emperor's intention of introducing this cult; it resembles more of a general phenomenon, independent of the political factor.[681] The vast majority of dedications are towards *genius centuriae,* in the legions and the praetorian cohorts indicating the men's attachment to their tactic and identity union. Most often the dedicators are *principales* of the *centuria* but also centurions and their sanctuaries are next to the barracks.[682]

The only institutionalized way of advancing from the base towards the top of the social pyramid is the centurionate career and, by promotion to *primus pilus,* admittance into the *ordo equester*. But, at the middle of the 2nd century there were about 2000 centurions of whom only a third advanced to *primus pilus* and the equestrian order and less than a tenth of them to a higher function in the equestrian career.[683]

It is not only the senatorial and equestrian officers who enjoy a privileged social status. Starting with Severus, from the rank of centurion and above, the soldiers make up a rather coherent social group, with political influence, prestige and a safe economic position. This social elevation of the military can best be observed in the offering of the golden ring, the statutory symbol of the *equites*,[684] to centurions and *principales* as proof of them being considered as potential knights. This '*esprit de corps*' can also be noticed from the banding of soldiers according to ranks in those associations *scholae* and *collegia*. Also they share the same cults and become proud of the politic role of the army because it is the same Severus who grants them the right to legal marriage,[685] thus they become active members of the urban society around the forts. Even more so, the inheritance of the military profession is another contribution to the solidification of this new social layer.[686]

---

671 Tacitus, *Ann.* 14.27.3. The fact that hierarchical relations were still respected even after discharge can be also deduced from AE 1975, 570, AE 1923, 33. MacMullen 1984, 442.

672 Caesar, *BG* 1.43, 7.17, *BC* 1.64. MacMullen 1984, 455.

673 The difference between promoted centurions and those directly appointed resided in their education. For example, a legionary centurion, problably *ex equite Romano*, addresses Cerealis on a familiar tone. *Tab. Vind.* 2.255.

674 Polybios 6.34.7-12.

675 The famous writer M. Valerius Probus tried for many years to become centurion and although he failed this example does not prove the expulsion of very educated men from the army. P. Alfenius Varus, a *trecenarius* makes a *chirographum*, indicating he knows how to write but not also his intellectual level. Vegetius, *Epit.* 2.19; Suetonius, *Gramm.* 24.1.

676 Not all centurions who speak two languages know Greek and Latin. The future emperor Maximinus Thrax, whilst still a young soldier serving under Severus, did not speak Latin and addresses the emperor in 'an almost Thracian' language but, later on, he becomes a centurion and *primipilus* and thus must have learnt Latin pretty well. *HA*, *Maxim.* 2.5. Another example of bilingualism is when a barbarian language is learnt as is the case of the centurion discovered in Slovakia, translator and *negotiator* with the *Quadi*. To conclude we can say that the education of the officers of the Roman army, particularly that of centurions, was just as heterogeneous as their social origin. Adams 1999, 132-133.

677 Richier 2004, 570.

678 Piret 1995, 237, Richier 2004, 560-561.

679 AE 1926, 69.

680 CIL III 1646.

681 Speidel, Dimitrova 1978, 1544-1545.

682 For example at Lambaesis. Speidel, Dimitrova 1978, 1546. An altar discovered at Novae, it is a dedication to Mars and the *genio armamentarii*, by a *custos armorum*. Although it is believed that the *armamentaria* belong to the main buildings of a fortification, it was found farther away and the connection with *custos armorum* is strange. In the meantime another such dedication was found, this time in the *principia* of the fortress from Potaissa, *genio armamentarii posuit Aelius Vitalis custos armorum*. Speidel, Dimitrova 1978, 1551-1552; Bărbulescu 2004, 376.

683 Alföldy 1979, 134.

684 *Digest* 3.8.5.

685 But not necessarily the right to live with his wife in the fortification, Hoffmann 1995, 110.

686 Alföldy 1979, 151-152.

II.2. Legionary centurions in Dacia

II.2.1. Centurions' origin and entry to the rank, in the army of Dacia[687]

There are over 40 centurions of ***XIII Gemina*** attested in Dacia. Furthermore, one was a centurion in this legion but also in the *V Macedonica*, which is why he was inserted in the statistical data of the other legion. We know the origin of roughly half of this number. Six are Italians,[688] eight are Eastern or Greeks,[689] anyway from the Eastern half of the Empire, four are from the African provinces,[690] two from Pannonia,[691] one each from Dalmatia,[692] Upper and Lower Moesia[693] and Hispania[694] and one of them bears a Celtic name.[695] As noticeable, the analysis of the situation of *XIII Gemina* alone is enough to draw some conclusions indicating as indubitable an increasing provincializing recruitment of legionary centurions in the 2nd-3rd centuries. Only 12% of the legion's centurions still have Italian origin, being surpassed in numbers even by those from the Eastern half of the Empir[696]e.

Out of the almost 60 centurions known to us from when the legion ***V Macedonica*** stationed in Dacia, we find details or mentions of their origin only for 15. Seven are Italians,[697] two come from Moesia Sup.,[698] respectively Sirmium, one from Dalmatia,[699] two seem to be Oriental-Greeks,[700] one bears a name of German resonance,[701] another of Illyrian one[702] and finally one has his name attested in Taracco.[703]

Of the eight centurions from the legion ***IIII Flavia*** mentioned in Dacia, the origin is mentioned or can be deduced from the inscriptions for only five of them. Three are Italians[704] (one might also be from Narbonensis), one, although presumably from Camulodunum, has been considered to be rather from Philippi,[705] Macedonia and the last one seems to be Oriental, judging by the inscription for *IOM Heliopolitanus*.[706]

As for the centurions coming from legions outside Dacia, the information is very sketchy. In the case of one centurion we can clearly establish the origin, Noricum,[707] while in the case of another there are several variants, one based on the Celtic origin of his name and the other three, Apulum, Sarmizegetusa or Nikopolis ad Istrum, based on the tribe.[708] Another one who belongs to a vexillation from *X Fretensis* that probably arrived here with the occasion of the Marcomannic Wars seems to be Eastern .[709]

[687] The tables in this section are marked as follows: **O** – origin of the centurion, **A** – career previous to the centurionate, **C** – career as centurion and **P** – career following the centurionate; they are numbered continuously in order to facilitate introducing references in the text.
[688] O2, O3, O4, O12 and O18.
[689] O1, O13, O14, O16, O17, O19, O20 and O22.
[690] O5, O6, O9 and O23.
[691] O7 and O10.
[692] O8.
[693] O11 and O15.
[694] O24.
[695] O21.
[696] Another Celer, Q. Attius, is also a centurion of *IIII Scythica*, like O1, and he is mentioned on the funerary stone of his slave, discovered in Western Anatolia. Bennett 2007, 405-406 considers his origin from somewhere in the West: Italy, Cisapline Gaul or Narbonensis. It is peculiar how the two bear the same *cognomen*, and are centurions of the same legion, about at the same moment in time. Perhaps Atilius Celer could also be considered as a local, therefore any efforts to explain the presence of a squadron of *IIII Scythica* in Anatolia would be useless.
[697] O25, O30, O31, O33, O34, O37 and O39.
[698] O26 and O29.
[699] O27.
[700] O28 and O32.
[701] O35. Another Turranius Severus, centurion of *XV Apollinaris*, is originary from Bellunum, Italy. Mosser 2003, 127. We can anyway consider him from the West, altogether with the *cognomen* leading to this conlusion.
[702] O36.
[703] O38.
[704] O40, O41 and O44.
[705] O42.
[706] O43.
[707] O46.
[708] O45.
[709] O47

O	Name	Origin	Date	Source	Place
		XIII Gemina			
1	C. Iul. Celer	Asia		CIL III 1044 = IDR III/5 148	Apulum
2	Silius Crispinus	Italy	AD 153-156	AE 1977, 653 = IDR III/5 166	Apulum
3	T. Pontius Sabinus	Italy	Trajan - Ant. Pius	CIL X 5829 = IDRE I 90	Ferentinum
4	M. Sabidius Maximus	Italy		AE 1937, 101 = IDRE II 364	Elbassan
5	C. Sulgius Caecilianus	Africa?	Severus Alex.	CIL VIII 14854 = IDRE II 431	Tuccabor
6	M. Iul. Quadratus	Numidia?		AE 1957, 249 = IDRE II 443	Castellum Arsacalitanum
7	M. Aebutius Victorinus	Pannonia		CIL III 6761 = IDRE II 399	Ancyra
8	Aequus	Dalmatia		CIL XIII 6952	Mogontiacum
9	Iul. Bassus Suplicianus	Numidia?	Severus Alex.	CIL VIII 2891 = IDRE II 455	Lambaesis
10	M. Aur. Lucius	Pannonia Sup.		AE 1941, 161 = IDRE II 404	Tyana
11	Aur. Domi(tius?)	Moesia Sup.?		AE 1938, 96 = IDRE II 314	Ratiaria
12	Burius Nocina	Italy		CIL V 6512 = IDRE I 164	Novaria
13	*Unknown*	Licia		CIL III 14178	Cyaneae
14	Ti. Cl. Vibianus	Phrygia		CIL III 7041	Augustopolis
15	L. Aemilius Severus	Moesia Inf.?		CIL III 14214(8)	Adamclisi
16	L. Licinius Messalinus	Eastern?		CIL III 1354 = IDR III/3 96	Micia
17	C. Domitius Valens	Eastern?		AE 1933, 18 = IDR III/2 243	Sarmizegetusa
18	M. Aur. Claudianus	Italy	After AD 150	AE 1981, 158 = IDRE I 98	Ostia
19	C. Iul. Lysias	Greek	Severus Alex.	CIL III 14469 = IDR III/5 32	Apulum
20	C. Cassius Proculeianus	Cilicia/Syria		AE 1960, 239 = IDR III/5 140	Apulum
21	M. Domestius Restitutus	Celtic	Aug. 23rd, AD 154	AE 1944, 28 = IDR III/5 232	Apulum
22	Archelaus	Greek		AE 1998, 1082 = IDR III/5 712	Apulum
23	C. Campanus	Africa	Before AD 238	AE 1916, 39 = IDRE II 428	Thuburbo Maius
24	C. Iul. Lepidus	Hispania		CIL II 4463 = IDRE I 174	Ieso
		V Macedonica			
25	Amblasius Secundus	Italy	Severan	CIL XI 710 = IDRE I 133	Bononia
26	M. Ulpius Titus	Moesia Sup.	3rd c. AD	CIL III 3259 = IDRE II 290	Cusum
27	L. Artorius Castus	Dalmatia	Antonines	CIL III 1919 = IDRE II 303	Epetium
28	Iulius Proculus	Eastern?		CIL VIII 2627	Lambaesis
29	Maecius Domitius	Moesia Sup.?	Antonines	IDR II 36	Drobeta
30	M. Pollius Hispanus	Italy	Antonines	AE 1992, 1470 =ILD 464	Potaissa
31	Trossius Crispus	Italy	Antonines	AE 1950, 15 = ILD 486	Potaissa
32	C. Iulius Antigonus	Greek/Eastern		CIL III 881	Potaissa
33	Cassius Severinus	Italy	Severan	AE 1971, 366 = IDR III/4 83a	Potaissa
34	Munatius Pius	Italy	Caracalla	ICLP 10-11	Potaissa
35	Turranius Germanus	Germany?	Caracalla	ICLP 10-11	Potaissa
36	Aelius Dardanus	Illyricum?	Caracalla	ICLP 10-11	Potaissa
37	Dovius	Italy?	Caracalla	ICLP 10-11	Potaissa
38	Aur. Tyrranus	Hispanic?	Caracalla	ICLP 10-11	Potaissa
39	Pupius	Italy?	Caracalla	ICLP 10-11	Potaissa
		IIII Flavia			
40	C. Caesius Silvester	Italy	2nd C.AD	CIL XI 5696 = IDRE I 126	Tuficum
41	Sex. Pilonius Modestus	Italy	Trajan	CIL III 1480 = IDR III/2 437	Sarmizegetusa
42	C. Titius Ianuarius	Philippi	Trajan	AE 1959, 314 = IDR II 35	Drobeta
43	Q. Licinius Macrinus	Eastern?	Trajan	CIL III 1353 = IDR III/3 95	Micia

44	L. Valerius Seranus	Italy/ Narbonensis	Trajan	IDR III/1 1	Kovin
		Legions outside Dacia			
45	P. Tenacius Vindex	Celtic/Dacia	Antonines	CIL III 1481, 6265 = IDR III/5 582, 518	Apulum
46	C. Censorinus Serenus	Noricum		CIL III 1615 = IDR III/5 513	Apulum
47	P. Caius Valerinus	Eastern?		CIL III 7625	Domnesti

Table II.2. Origins of legionary centurions from Dacia

The entry to the centurionate. When compared to other legions on the aspect of previous careers of its centurions, ***XIII Gemina*** fares better, with eight centurions giving such details. Four of them come from the legionary *caligati*, one detailing his career as a *principalis* in *XI Claudia – miles, signifer, cornicularius, optio ad spem ordinis*.[710] Three other mention one post, possibly the one they were promoted from to centurion, respectively *strator legati Augusti pro praetore*,[711] *aquilifer*[712] and *speculator*.[713] One of them has served as soldier in Rome, even though not in praetorian cohorts. He originates from and is specialized in the military navy. He starts his career as *optio peregrinorum*, then is promoted in the *castra peregrina*, also being appointed here *exercitator militum frumentariorum* and becomes a fleet officer in the praetorian fleet, *navarchus classis Missenatium*, post from where he will advance to centurion.[714] Finally, two knights are mentioned :[715] one of them breaks off with his equestrian career that he had already started after completing two equestrian militias .[716]

In the legion ***V Macedonica*** we can ascertain such information only for four centurions: two of them were legionaries, going through the *officium consularis* of Dacia, on the post of *speculator*,[717] one was a praetorian with a military career that includes five posts: *beneficiarius tribuni, tesserarius, optio, signifer, beneficiarius praefecti praetorio, evocatus Augusti*,[718] taking up afterwards all the three Roman centurionates and the title *trecenarius*. Finally, a former praetorian centurion comes in this legion probably to occupy the primipilate.[719] The centurion lists in Potaissa presents him only as a *primus ordo, I princeps posterior*. In the case of perhaps only one centurion, the one from Dalmatia, can we assume membership in the equestrian order?

As far as access to the centurionate in the legion ***IIII Flavia*** we have information concerning only two centurions: one is a knight and joins the army directly appointed centurion[720] and the other was a praetorian who mentions only two previously held posts before the centurionate – probably the most important ones: *beneficiarius praefecti praetorio* and *evocatus Augusti*.[721]

There seems to be a single centurion originating from a legion outside Dacia who mentions a post held before the centurionate, that of *eques legionis*, but his affiliation is not certain .[722]

---

[710] A4.
[711] A1.
[712] A7.
[713] A2.
[714] A5.
[715] A3 and A6.
[716] A3.
[717] A9 and A10.
[718] A8.
[719] A11.
[720] A13.
[721] A12.
[722] A14.

A	Name	Previous career	Date	Source	Place
		XIII Gemina			
1	Silius Crispinus	*strator leg Aug pr pr*	AD 153-156	AE 1977, 653 = IDR III/5 166	Apulum
2	Ulpius Bacchius	*speculator*	AD 198-199	CIL III 14479 = IDR III/5 426	Apulum
3	T. Pontius Sabinus	*praef coh I Pann et Dalm eq c R*	Trajan - Ant. Pius	CIL X 5829 = IDRE I 90	Ferentinum
		trib leg VI Ferratae			
4	M. Sabidius Maximus	*miles leg XI Claudiae*		AE 1937, 101 = IDRE II 364	Elbassan
		signifer			
		cornicularius			
		optio ad spem ordinis			
5	C. Sulgius Caecilianus	*optio peregrinorum*	Elagabal- Severus Alex.	CIL VIII 14854	Tuccabor
		exerc militum frum navarchus cls Misen			
6	M. Iulius Quadratus	*eques publicus*		AE 1957, 249 = IDRE II 443	Castellum Arsacalitanum
7	Aequus ?	*aquilifer leg XXX Ulp victr*		CIL XIII 6952	Mogontiacum
		V Macedonica			
8	Amblasius Secundus	*miles coh I (praet)*	after Severus	CIL XI 710 = IDRE I 133	Bononia
		beneficiarius tribuni coh I (praet)			
		tesserarius			
		optio			
		signifer			
		beneficiarius praef praet			
		evocatus Aug			
		> coh I vigilum			
		> coh XIIII urbanae			
		> coh ? praetoriae			
9	Iulius Tacitus	*speculator*	AD 198-199	IDR III/5 426	Apulum
10	Cl. Claudianus	*speculator*	AD 198-199	IDR III/5 426	Apulum
11	Munatius Pius	*> coh I praet*	AD 212-220	IDRE I 43=ICLP 10-11	Apulum
		IIII Flavia			
12	C. Caesius Silvester	*beneficiarius praef praet*	2nd c. AD	CIL XI 5696 = IDRE I 126	Tuficum
		evocatus Aug			
13	Sex. Pilonius Modestus	*ex equite Romano*	Trajan	CIL III 1480 = IDR III/2 437	Sarmizegetusa
		Other legions			
14	Ferox	*eques legionis*		CIL III 8048 = IDR II 644	Romula?

Table II.3. Access to the centurionate in Dacia.

II.2.2. Centurion careers in Dacia

A number of 14 centurions from ***XIII Gemina*** mention transfers or centurionates held in other legions. The number of transfers varies between two and ten, five of them already having three transfers. Five centurions mention their exact position in the legion: *X hastatus posterior, I hastatus posterior, II princeps prior, hastatus prior, hastatus* and *princeps*.[723] Unfortunately we cannot come to any conclusion, except for the knight who holds only two centurion posts prior to becoming *primus pilus*, the second being *princeps legionis XIII Geminae*, confirming both the hierarchy of the first three centurions from a legion as well as a certain advancement rule.[724] Three centurions are mentioned holding an extraordinary function: one is *exercitator equitum singularium*,[725] the other two being *curam agentes* of several vexillations tasked with construction work, one probably at the fortification from Hoghiz, the other with reconstruction the *tabularium*.[726] For three centurions we also have chronological information regarding their careers. The centurion with the career of *principalis* in the *XI Claudia* was a soldier for 20 years and held 7 posts as centurion in the remaining 20 years of his service .[727] The character with eight mentioned centurionates held them in just 14 years of service ,[728] the opposite case being an *aquilifer* recruited at 25 years of age with only three posts of centurion and 45 years of service .[729]

[723] C42, C27, C13, C23, C6 and C8.
[724] C8.
[725] C5.
[726] C1 and C41.
[727] C9.
[728] C12.
[729] C14.

C	Name	Centurionates	Other posts	Date	Source	Place
1	Antoninianus	*XIII Gemina*	*curam agens vexill*	Hadrian	CIL III 953 = IDR III/4 230	Hoghiz
2	Ti. Claudius Valerian.	*XIII Gemina* *I Adiutrix*		Trajan	CIL III 981 = IDR III/5 11	Apulum
3	C. Iulius Celer	*IIII Scythica* *XVI Fl firma* *XIII Gemina*			CIL III 1044 = IDR III/5 148	Apulum
4	Silius Crispinus	*XIII Gemina*		AD 153-156	AE 1977, 653=IDR III/5 166	Apulum
5	M. Ulpius Restitutus	*XIII Gemina*	*exerc eq singularium*		CIL III 14477=IDR III/5 401	Apulum
6	*Ignotus*	*XIII Gemina*	*hastatus*		CIL III 14477=IDR III/5 401	Apulum
7	Ulpius Bacchius	*XIII Gemina*		AD 198-199	CIL III 14479=IDR III/5 426	Apulum
8	T. Pontius Sabinus	*XXII Primig* *XIII Gemina*	*hastatus* *princeps*	Trajan - Ant. Pius	CIL X 5829 = IDRE I 90	Ferentinum
9	M. Sabidius Maximus	*XI Claudia* *III Gallica* *IIII Scythica* *3 x ?* *XIII Gemina*	*stip cent 20* *stip continua 40*		AE 1937, 101 = IDRE II 364	Elbassan
10	C. Sulgius Caecilianus	*XIII Gemina* *XVI Fl firma* *I Parthica* *VII Gemina* *III Augusta*	*praep reliquationis* *classis Missenatium*	Elagabal-Sev. Alex.	CIL VIII 14854=IDRE II 431	Tuccabor
11	M. Iulius Quadratus	*XIII Gemina* *III Augusta* *II Augusta*			AE 1957, 249 = IDRE II 443	Castellum Arsacalitanum
12	M. Aebutius Victorin.	*X Gemina* *XI Claudia* *XIIII Gemina* *I Minervia* *XXII Primig* *XIII Gemina* *VII Claudia* *XV Apollinaris*	*stip 14*		CIL III 6761 = IDRE II 399	Ancyra
13	M. Aurelius Claudianus	*XXII Primig* *XV Apollinaris* *I Minervia* *VII Claudia* *III Gallica* *XIII Gemina* *IIII Flavia* *I Italica bis* *II Adiutrix* *XVI Fl firma*	*II princeps prior*	after AD 150	AE 1981, 158 = IDRE I 98	Ostia
14	Aequus ?	*XIIII Gemina*	*stip 45*		CIL XIII 6952	Mogontiacum

		XIII Gemina *XXII Primig*				
15	Iul. Bassus Suplicianus	*II Traiana* *XXII Primig* *XIII Gemina* *III Augusta* *III Parthica*		Sev. Alexander	CIL VIII 2891 = IDRE II 455	Lambaesis
16	Avidius Valens	*XIII Gemina*		AD 170	CIL VIII 9365 = IDRE II 459	Caesarea (Maur.)
17	M. Aurelius Lucius	*XIII Gemina*			AE 1941, 161 = IDRE II 404	Tyana
18	C. Iulius Placideius	*XIII Gemina*			AE 1939, 61 = IDRE II 408	Heliopolis
19	Spurius Maximus	*XIII Gemina*			AE 1928, 84 = IDRE II 409	Beyrouth
20	Marcius Vitalis	*XIII Gemina*			AE 1938, 916 = IDRE II 379	Ephesus
21	Aurelius Domi(tius?)	*XIII Gemina*			AE 1938, 96 = IDRE II 314	Ratiaria
22	Burius Nocina	*XIII Gemina*			CIL V 6512 = IDRE I 164	Novaria
23	*Ignotus*	*XI Claudia* *XIII Gemina* *IIII Scythica*	*hastatus prior*		CIL XIII 1859	Lugdunum
24	*Ignotus*	*XIII Gemina* *XV Apollinaris* *XXX Ulpia v.*			CIL III 14178	Cyaneae
25	Ti. Claudius Vibianus	*XIII Gemina*	*> frumentarius*		CIL III 7041=IDRE II 385	Augustopolis
26	L. Aemilius Severus	*XIII Gemina*			CIL III 14214(8)	Adamclisi
27	C. Iulius Certus	*XIII Gemina*	*I hastatus posterior*		CIL VI 31159	Roma
28	Claudius Marus	*XIII Gemina*			IDR I Tcer VIII	Alburnus Maior
29	L. Licinius Messalinus	*XIII Gemina*			CIL III 1354 = IDR III/3 96	Micia
30	C. Iulius Iulianus	*XIII Gemina*			CIL III 7858 = IDR III/3 115	Micia
31	*III principes*	*XIII Gemina*			IDR III/3 260 = ILD 316	Orăştioara de Sus
32	Siscius Valerius	*XIII Gemina*		AD 244-249	CIL III 1464 = IDR III/2 100	Sarmizegetusa
33	C. Domitius Valens	*XIII Gemina*			AE 1933, 18 = IDR III/2 243	Sarmizegetusa
34	C. Laberius Iustus	*XIII Gemina*			IDR III/2 248	Sarmizegetusa
35	C. Iavolenus Modestus	*XIII Gemina*			CIL III 1434 = IDR III/2 268	Sarmizegetusa
36	C. Iulius Macer	*XIII Gemina*			CIL III 1478 = IDR III/2 428	Sarmizegetusa
37	C. Iulius Lysias	*XIII Gemina*		Sev. Alexander	CIL III 14469 = IDR III/5 32	Apulum
38	C. Cassius Proculeianus	*XIII Gemina*			AE 1960, 239=IDR III/5 140	Apulum
39	Florius Maximus	*XIII Gemina*			IDR III/5 146	Apulum
40	Iulius Primus	*XIII Gemina*			CIL III 1045 = IDR III/5 152	Apulum
41	Primus Senecio	*XIII Gemina*	*curam ag tabularii*		AE 1971, 369=IDR III/5 205	Apulum
42	M. Domestius Restitut	*XIII Gemina*	*X hastatus post*	Aug. 23rd, AD 154	AE 1944, 28 = IDR III/5 232	Apulum
43	Archelaus	*XIII Gemina*			AE 1998,1082=IDR III/5 712	Apulum
44	C. Campanus	*XIII Gemina*		before AD 238	AE 191, 39 = IRDE II 428	Thuburbo Maius
45	C. Iul. Lepidus	*XIII Gemina* *I Adiutrix* *X et VII Geminae* *XX Valeria V*			CIL II 4463=IDRE I 174	Ieso

Table. II.4.1. Centurion careers in *XIII Gemina*.

Very few centurions from the legion ***V Macedonica*** offer details regarding their service as centurions. The former praetorian, after being recalled for service, holds consecutively the three centurionate posts in Rome and also receive the title *trecenarius*, then follows a single centurionate in the *V Macedonica* before the primpilate .[730] Three other centurions mention transfers between several legions: between three and six.[731] Four centurions from *V Macedonica* also occupy extraordinary functions. Two of them are *praepositi* for two *numeri* newly created in Dacia[732] while a third supervises the reconstruction of Bel's temple from Porolissum – apparently done by an auxiliary construction vexillation – from his post of *curam agens* .[733] No details regarding posts or hierarchy are given. Finally, one is detached with a vexillation of the legion in Dobrudja, Moesia Inferior, as *curam agens*, together with a squadron of *I Italica*, having also a centurion in charge.[734]

To these we should add the recently published double dedication to Caracalla and Julia Domna, made by the centurions of *V Macedonica*, in the *principia* of the fortress of Potaissa.[735]

The first two rows contain the name and imperial titles; the third begins with *LEG V MAC ANTONINIANAE DEVOTI NVM EOR*. The end of the second row poses a problem because of the presence of *VNIVERS(A/I?)*. The two options, *universa legio V...*[736] and *centuriones universi legionis V...*[737] both incline to tell that the entire legion was meant to assist in these dedications.
They are followed by the list of centurions present at that time in the legion. There are 38 (39) names from which three are supposed and at least eight are undecipherable. They are divided by cohort starting with the first, disposed on five columns, but in regard of rows, no logical layout can be deducted. This is why we cannot say exactly how many undiscovered names would fit in the empty spaces.[738]

The only cohort that is at full strength seems to be the first. It has all its five centurions. The *primus pilus* is simply Tib(erius/urtius?[739]), not encountered before, and we have reason to believe that the following order is that of the hierarchy of the *primi ordines*: *princeps*, *hastatus*, *princeps posterior*, *hastatus posterior*. The name of the *princeps posterior* seems to be quite rare and encountered as a centurion in *I praetoria*, during Severus' reign.[740] It could be the same character, confirming once again that praetorian centurions come into legions directly as *primi ordines*, expecting the primipilate soon.

We could extend the order principle to all the cohorts, but as they are not at full strength, we don't actually know which *centuriae* are left. All the other centurions are not attested elsewhere. Of course there are three centurions of the same legion called Saturninus,[741] Severinus[742] and Maximinus,[743] but to equate them only on this argument could be hazardous. Therefore not much can be said about their origin or careers. They are all citizens, as we could expect, but some of them rather recently made, due to the name Aurelius. I believe that at this time one can assume that provincial recruitment has extended also to legionary centurions, the only 'intruders' being either men coming from Rome or other beneficiaries of favors.

If no other centurion is to be presumed, we have therefore only 39 present at this time in Potaissa and of course in service, because no centurion in service would afford to miss out on such a dedication. Some conclusions can be drawn about the legion's strength. As I said, the first cohort is in full strength, maybe not for military reasons, but only for the fact that the *primi ordines* spots were the first to fill, in the expectation of the primipilate, therefore 960 men. The 2nd has only two centurions, therefore we would assume 160 men; the 3rd has three or four, therefore 400 men at best; the 4th has three, therefore 240 men; the 5th has the same, 240; the 6th has only two, therefore 160 men; the 7th could have five, so 400 men at best, the 8th four centurions, that make for 320 men; the 9th maybe four, again 320 men and finally the 10th five at maximum, for 400 men.

The total would give us 3,600 legionnaires in the *V Macedonica* somewhere during Caracalla's reign. This counts for 68% of the full strength of the legion. What happened to the other 32%? I. Piso proposed the 'knock-on effect' in the displacement of the legions, especially at the end of the 2nd century and the beginning of the 3rd. This means that when part of a legion leaves in a campaign, its place is taken by squadrons coming from other legions. In this case, many Romanian scholars propose the *VII Gemina*, from Léon, because it presents evidence of work and activity in Dacia Porolissensis, at about this time (several tile stamps at Porolissum[744] and the centurions erecting a monument for their comrades at Potaissa[745]).[746]

Another option of interpretation is that the legion in Potaissa never had full strength. This can also be confirmed by the weak state of the constructions inside the fortress, except the official buildings and the baths. Barracks are almost inexistent, and the fortress has several large empty spaces.[747]

[730] C46.
[731] C52, C53 and C54.
[732] C47 and C48.
[733] C51.
[734] C57.
[735] One of the fragments is CIL III 922, the others gathered in Bărbulescu 2012, nos. 10-11.
[736] Bărbulescu 2012, 113 sqq.
[737] Suggested to me by Prof. I. Piso, to whom I am grateful.
[738] C69-C105.
[739] Another Tiburtius is centurion of *XXX Ulpia victrix*, but he dies at Xanten. Reuter 2012, 94-95.
[740] Bărbulescu 2012, 121. (CIL VI 32533, 32640).
[741] C85=C51?
[742] C95=C68?
[743] C96=C47?
[744] See Gudea 1976, 109-113.
[745] C63 and C117.
[746] Piso *et al* 2012, 72-73.
[747] For details see Barbulescu 1987 and Barbulescu 1997.

C	Name	Centurionates	Other posts	Date	Source	Place
46	Amblasius Secundus	*coh I vigilum*	*ex CCC*	after Severus	CIL XI 710 = IDRE I 133	Bononia
		coh XIIII urbana				
		coh ? praetoria				
		V Macedonica				
47	Ulpius Maximinus	*V Macedonica*	*praep n sing ped Britann*	AD 238-244	AE 1967, 410=IDR III/3 237	Germisara
48	M. Verius Superstes	*V Macedonica*	*praep n Germanorum*	AD 176-177	CIL III 12574=IDR III/3 262	Orăştioara de Sus
49	Iulius Tacitus	*V Macedonica*		AD 198-199	CIL III 14479=IDR III/5 426	Apulum
50	Claudius Claudianus	*V Macedonica*		AD 198-199	CIL III 14479=IDR III/5 426	Apulum
51	T. Flavius Saturninus	*V Macedonica*	*curam ag n Palmyr sagit*	Caracalla	AE 1980, 755=ILD 663	Porolissum
52	M. Ulpius Titus	*II Adiutrix*		3rd c. AD	CIL III 3259=IDRE II 290	Cusum
		I Adiutrix				
		XIII Gemina				
		V Macedonica				
		XIIII Gemina				
		II Adiutrix				
53	L. Artorius Castus	*III Gallica*		M. Aur. - Commodus	CIL III 1919=IDRE II 303	Epetium
		VI Ferrata			CIL III 12224=IDRE II 304	Salonae
		II Adiutrix				
		V Macedonica				
54	Iulius Proculus	*V Macedonica*			CIL VIII 2627	Lambaesis
		III Gallica				
		XXII Primigenia				
55	Maecius Domitius	*V Macedonica*		Antonines	IDR II 36	Drobeta
56	Ti. Cl. Ambrelianus	*V Macedonica*		Severus	CIL VI 1408=IDRE I 12	Rome
57	Epitidius Modestus	*V Macedonica*	*curam agens vexill*		CIL III 14433	Ghiuvegea
58	Cassius Severus	*V Macedonica*			IDR III/4 83a = ILD 473	Potaissa?
59	Q. Cornelius Silvanus	*V Macedonica*		Commodus _	CIL III 7764=IDR III/5 197	Apulum
60	M. Pollius Hispanus	*V Macedonica*		Commodus _	AE 1992, 1470 = ILD 464	Potaissa
61	C. Cassius Vitalis	*V Macedonica*		M. Aurelius	AE 1972, 454 = ILD 482	Potaissa
62	Trossius Crispus	*V Macedonica*		Commodus _	AE 1950, 15 = ILD 486	Potaissa
63	C. Valerius Lucanus	*V Macedonica*		Commodus _	AE 1976, 574 = ILD 499	Potaissa
64	Valerius Iustinus	*V Macedonica*			CIL III 7672=ILD 502	Potaissa
65	C. Iulius Antigonus	*V Macedonica*			CIL III 881	Potaissa
66	Serennus Martinianus	*V Macedonica*			CIL III 7692	Potaissa
67	Aurelius C(?)	*V Macedonica*			CIL III 1603	Potaissa?
68	Cassius Severinus	*V Macedonica*			AE 1971, 366=IDR III/4 83a	Lunca Muresului
69	Terentius Vibianus	*V Macedonica*	*princeps*		ICLP 10-11	Potaissa
70	Provincialis Rufus	*V Macedonica*	*hastatus*		ICLP 10-11	Potaissa
71	Munatius Pius	*V Macedonica*	*I princeps posterior*		ICLP 10-11	Potaissa
72	A(?)	*V Macedonica*	*I hastatus posterior*		ICLP 10-11	Potaissa
73	Ael. Aelianus	*V Macedonica*	*II*		ICLP 10-11	Potaissa
74	Cl. Valentinus	*V Macedonica*	*II*		ICLP 10-11	Potaissa

75	Valerius Antonius	*V Macedonica*	*III*	ICLP 10-11	Potaissa
76	Domitius	*V Macedonica*	*III*	ICLP 10-11	Potaissa
77	Frue(ius?)	*V Macedonica*	*III*	ICLP 10-11	Potaissa
78	Septimius	*V Macedonica*	*III*	ICLP 10-11	Potaissa
79	*Unknown*	*V Macedonica*	*III*	ICLP 10-11	Potaissa
80	Cocceius Dignus	*V Macedonica*	*IIII*	ICLP 10-11	Potaissa
81	Turranius Germanus	*V Macedonica*	*IIII*	ICLP 10-11	Potaissa
82	Ael. Dardanus	*V Macedonica*	*IIII*	ICLP 10-11	Potaissa
83	Aur. Ulpius	*V Macedonica*	*V*	ICLP 10-11	Potaissa
84	Aur. Annius	*V Macedonica*	*V*	ICLP 10-11	Potaissa
85	Saturninus = 51?	*V Macedonica*	*V*	ICLP 10-11	Potaissa
86	Valens	*V Macedonica*	*VI*	ICLP 10-11	Potaissa
87	*Unknown*	*V Macedonica*	*VI*	ICLP 10-11	Potaissa
88	Dovius	*V Macedonica*	*VII*	ICLP 10-11	Potaissa
89	Servilius	*V Macedonica*	*VII*	ICLP 10-11	Potaissa
90	*Unknown*	*V Macedonica*	*VII?*	ICLP 10-11	Potaissa
91	*Unknown*	*V Macedonica*	*VII?*	ICLP 10-11	Potaissa
92	*Unknown*	*V Macedonica*	*VII?*	ICLP 10-11	Potaissa
93	Aur. Maximus	*V Macedonica*	*VIII*	ICLP 10-11	Potaissa
94	Niger	*V Macedonica*	*VIII*	ICLP 10-11	Potaissa
95	Severinus = 68?	*V Macedonica*	*VIII*	ICLP 10-11	Potaissa
96	Maximinus = 47?	*V Macedonica*	*VIII*	ICLP 10-11	Potaissa
97	Aur. Tyrannus	*V Macedonica*	*VIIII?*	ICLP 10-11	Potaissa
98	Aur. Vitalis	*V Macedonica*	*VIIII?*	ICLP 10-11	Potaissa
99	*Unknown*	*V Macedonica*	*VIIII*	ICLP 10-11	Potaissa
100	(Su?)ccessus	*V Macedonica*	*VIIII*	ICLP 10-11	Potaissa
101	*Unknown*	*V Macedonica*	*X*	ICLP 10-11	Potaissa
102	Aurelius	*V Macedonica*	*X*	ICLP 10-11	Potaissa
103	Pupius	*V Macedonica*	*X*	ICLP 10-11	Potaissa
104	Fl. Iu(stinus?)	*V Macedonica*	*X*	ICLP 10-11	Potaissa
105	Claudius	*V Macedonica*	*X*	ICLP 10-11	Potaissa
106	Faustinus	*V Macedonica*		ICLP 63	Potaissa
107	Maternus	*V Macedonica*		ICLP 64	Potaissa

Table. II.4.2. Centurion careers in *V Macedonica*

Four of the centurions from ***IIII Flavia*** give details about their service as centurions. Three of them have served in more than one legion, the number of transfer varying between two and five .[748] During his centurionate one is sent at the governor's headquarters to train his mounted guard - *exercitator equitum singularium*[749] and another seems to have held the post of *centurio regionarius*,[750] though he can also be considered *centurio frumentarius*.[751] As for details regarding the centurionate service, they are mentioned in only one case: that of the centurion transferred in five legions and who ended his career as *III hastatus posterior*.[752]

What about transfers at home? Is there any policy on rotation of the cadre inside the provincial army? In Dacia only few such examples can be observed, most of them involving *XIII Gemina*. Two of the transfers to the next centurionate in *I Adiutrix* can be dated in the time that both of the legions (or just a vexillation of the second) were garrisoned in Dacia, the reign of Trajan, or exactly during the Dacian Wars.[753] One of the examples of a transfer to *IIII Flavia* dates from after AD 150, so it cannot be considered as a transfer inside the provincial army.[754] *IIII Flavia* is involved in another example, that implies a transfer in *V Macedonica*, but the career is very hard to date exactly, and we cannot argue if this transfer occurred in Dacia.[755] Finally, *V Macedonica* receives a centurion from *XIII Gemina*, at the beginning of the 3rd century, this being the only clear example of a centurion transferred between the legions of the same province in Dacia.[756]

We find chronologic details regarding the careers in only one case. He dies at the age of 37, after serving 19 years exclusively as centurion in five legions .[757] It is the soldier who became centurion *ex equite Romano*, joining the army at the age of 18. This knight from Beneventum is enlisted very early. Under no circumstances was he interested in an equestrian career, seeking to become centurion from his early maturity. For a Roman knight he is one of the youngest to directly ascend to the centurionate, this perhaps being an explanation why he did not get to the primpilate. His mention, *ordinem accepit ex equite Romano*, is again proof for retaining his status as a knight while serving in the army, indicating only his social origin.[758] His service as a centurion, which lasted 19 years, takes place in five legions, the five transfers probably taking place at the time of Trajan's Dacian wars or perhaps immediately after the conquest, when all these legions were nearby. The last post held by Modestus is *III hastatus posterior*, not confirming any promotion system.

It is known that because of the threat of an invasion of the Carps, emperor Philippus ordered the fortification of Romula, a job for which building vexillations from at least two legions were dispatched. Although these vexillations do not mention their commanders they 'sign' building blocks or plaques. Thus, from the legion *VII Claudia* from Viminacium we find attested the *centuriae* commanded by *VIII hastati priores et posteriores* and *VIIII hastatus prior*.[759] From the *XXII Primigenia*, took part the *centuriae V hastati priores et posteriores* and *IIII pilus posterior*.[760] Three of the inscriptions are dated AD 248 and the mention of *centuria* after its place in the legion and not after the centurion's name corresponds with the tendencies noted by M.P. Speidel for the beginning of the 3rd century .[761] In the case of the *centuria VIIII hastati prioris* of *VII Claudia*, the building contingent seems to have been led by the *optio* and the *signifer*, who are mentioned on the inscription. Besides this, in the case of the *IIII pilus posterior* from *XXII Primigenia*, its mention is done by the abbreviation with a centurion sign - Γ.[762]

In principle, each of the centurions from the legions external to Dacia mention a single legion and only one offers chronological details. We are referring to the former legionary horseman who served a total of 25 years, 18 of them as *eques legionis* and died at the age of 43. Sometimes during his career, perhaps after his 18 years of mounted service, he became centurion .[763]

748 C108, C109 and C110.
749 C109.
750 C111. Speidel 1984, 186-187 and footnote no. 8.
751 The authors of IDR II, respectively ILD.
752 C110.
753 C2 and C45.
754 C13.
755 C109. For this career see Speidel 1994, Petolescu 2005 and Cupcea 2013a.
756 C52.
757 C110.
758 *Contra*, the authors of IDR III/2, who translate "was made centurion, former Roman knight".
759 C118 and C119.
760 C120 and C121. It is in this context that the epithet *vestra* for the legion *VII Claudia*, characteristic for the period of the Dominate, appears; see Speidel 1978a, 121.
761 Speidel 1978a, Speidel 1983, Speidel 1986a and Speidel 2005.
762 Speidel 1983 and Faure 2008, 301.
763 C126.

C	Name	Centurionates	Other posts	Date	Source	Place
108	C. Caesius Silvester	*II Augusta*		Mid 2nd c. AD	CIL XI 5696 = IDRE I 126	Tuficum
		IIII Flavia				
		III Gallica				
		VI Ferrata				
		XXX Ulpia victrix				
109	M. Calventius Viator	*IIII Flavia*	*exercitator eq sing*	Trajan-Hadrian	CIL III 7904 = IDR III/2 205	Sarmizegetusa
		V Macedonica			AE 1915, 42	Gerasa
					CIL VIII 2532=18042	Lambaesis
110	Sex. Pilonius Modestus	*VII Claudia*	*stip cent 19*	Trajan	CIL III 1480 = IDR III/2 437	Sarmizegetusa
		VIII Augusta	*III hastatus posterior*			
		Claudia				
		I Minervia				
		IIII Flavia				
111	C. Titius Ianuarius	*IIII Flavia*	*centurio regionarius*	Trajan	AE 1959, 314 = IDR II 35	Drobeta
112	Q. Licinius Macrinus	*IIII Flavia*		Trajan	CIL III 1353 = IDR III/3 95	Micia
113	L. Valerius Seranus	*IIII Flavia*		Trajan	IDR III/1 1	Kovin
114	Iulius Strato	*IIII Flavia*			CIL III 14484 = IDR II 34	Drobeta
			Other legions			
115	M. Ulpius Caius	*III Italica*		AD 213-215	CIL III 1178 = IDR III/5 436	Apulum
					IDR III/5 448	Apulum
116	Ulpius Vitalis	*III Italica*		After AD 168	CIL III 7785 = IDR III/5 402	Apulum
117	C. Tiberius Celer	*VII Claudia*			AE 1976, 574 = ILD 499	Potaissa
118	*VIII hast pr et post*	*VII Claudia*	*VIII hast pr et post*	Philippus	AE 1939, 28 = IDR II 327	Romula
119	*VIIII hast pr*	*VII Claudia*	*VIIII hastatus prior*	AD 248	CIL III 8074 = IDR II 328	Romula
120	*V hast pr et post*	*XXII Primigenia*	*V hast pr et post*	AD 248	AE 1940, 13 = IDR II 325	Romula
121	*IIII pilus post*	*XXII Primigenia*	*IIII pilus posterior*	AD 248	AE 1940, 14 = IDR II 326	Romula
122	P. Tenacius Vindex	*XXII Primigenia*		Commodus	CIL III 1481 = IDR III/2 120 = IDR III/5 582	Apulum
					CIL III 6265 = IDR III/5 518	Apulum
123	T. Aurelius Emeritus	*VI Victrix*		AD 222-235	AE 1933, 247 = IDR III/2 124, 127	Sarmizegetusa
124	C. Censorinus Serenus	*XIIII Gemina*			CIL III 1615 = IDR III/5 513	Apulum
125	P. Caius Valerinus	*X Fretensis*	*centurio regionarius*		CIL III 7625	Domneşti
126	Ferox	*?*	*stip continua 25*		CIL III 8048 = IDR II 644	Romula?
			stip eq leg 18			Sucidava?

Table II.4.3-4. Centurion careers in *IIII Flavia* and other legions.

II.2.3. Post-centurionate careers in Dacia

Surprisingly, only in the case of three centurions from the legion ***XIII Gemina*** attested in Dacia do we have information regarding a post-centurionate career. The knight with two equestrian militias takes part in Trajan's Parthian expedition as *tribunus legionis VI Ferratae* and he is decorated as an officer of equestrian rank. At this point he interrupts his equestrian career and, after serving twice as centurion, becomes *primipilus legionis III Augustae*. His next post is *praepositus vexillationibus milliaris legionum VII Geminae, VIII Augustae, XXII Primigeniae*, in an expedition from Britannia, thus a commander of an army corps made up of three-*milliariae* legionary vexillations coming from Hispania and Germania Superior. Next he continues uninterrupted with the three tribune posts from Rome: *tribunus cohortis III vigilum, XIIII urbanae* and *II praetoriae*, and then the post of *primipilus II*. This exceptional career opens his way to the post of *procurator ducenarius* for the province Narbonensis and multiple municipal and provincial honors.[764] This is one of the most beautiful centurionate careers from the 2nd century AD and it leaves us wondering whether it was not more profitable for a knight skilled in military affairs to change an equestrian career for a military one.

The second example of a centurion from the *XIII Gemina* with a post-centurionate career it that of the former *navarchus* and *optio peregrinorum* who, after five centurionate posts, the first one in the *XIII Gemina*, receives an extraordinary command as *praepositus reliquationi classis praetoriae Misenatium*, used for transporting men and resources for Severus Alexander's Parthian expedition. Then he becomes *primipilus legionis XX Valeriae victricis* and finally *praefectus legionis III Cyrenaicae*.[765]

Finally, the Spaniard Lepidus, after probably beginning his centurion career in the legion in Dacia, he passes through other four legionary centurionates, in Pannonia, Britain and Tarraconensis, finally he holds the primipilate.[766] As to where he is a primus pilus, it is hard to say, depending on the order of the centurionates on the monument. This could very well be also *XIII Gemina*.[767]

As far as post-centurionate careers are concerned, only two centurions from the ***V Macedonica*** mention them, but neither of them is attested in Dacia. The former praetorian and *trecenarius* is believed to have held a post of *primus pilus* in *I Adiutrix*.[768] The would-be knight from Dalmatia has an equestrian career including several posts typical of the military navy, probably an indication if his specialization. After being a *primus pilus* in *V Macedonica*, he receives an extraordinary command, as *praepositus classis Missenatium*, which seems to replace or be the equivalent of the position of tribune of the cohorts from Rome, nowhere mentioned in his career. Next comes the position of *praefectus* of the legion *VI Victrix*, in Britannia and again an extraordinary command as *dux leggionum cohortium alarum Britannicianarum adversus Armoricanos*, and finally a procuratorship of a region from Dalmatia, *procurator Liburniae iure gladii*, probably for fighting against piracy or brigandage.[769]

Two of the centurions from ***IIII Flavia*** advance in their careers after the centurionate. The former praetorian became *primus pilus* and *praefectus castrorum* of the legion *IIII Flavia*,[770] and another used his skill to train and even lead the riders in the guard of Emperor Hadrian – *curam agens equitum singularium Augusti*.[771]

As far as centurions from legions outside Dacia are concerned we can assume a further career in but only one case: that of a centurion from *III Italica* who, once he arrived in Dacia with a vexillation, might have been transferred in *XIII Gemina* and, following that, became praefect of this legion and of *IIII Flavia*.[772] Still, it can be a coincidence of names between two different characters.

[764] P1.
[765] P2.
[766] P3.
[767] Malone 2006, 119, no. 37.
[768] P4.
[769] P5. Petolescu, IDRE II 303.
[770] P6. If the centurionate is held while the legion is garrisoned in Dacia, the primipilate seems to be held at Singidunum. Reuter 2012, 81.
[771] P7.
[772] P8.

P	Name	Career	Date	Source	Place
		XIII Gemina			
1	T. Pontius Sabinus	*pp III Augustae* *praep vexill leg* *VII Geminae* *VIII Augustae* *XXII Primigeniae* *trib coh III vig* *trib coh XIIII urb* *trib coh II praet* *pp II* *proc Narbonensis*	Trajan-Pius	CIL X 5829=IDRE I 90	Ferentinum
2	C. Sulgius Caecilianus	*pp XX Val victr* *praef leg III Cyr*	Sev. Alexander	CIL VIII 14854=IDRE II 431	Tuccabor
3	C. Iul. Lepidus	*pp VII Geminae*		CIL II 446=IDRE I 174	Ieso
		V Macedonica			
4	Amblasius Secundus	*pp I Adiutricis ?*	after Severus	CIL XI 710=IDRE I 133	Bononia
5	L. Artorius Castus	*pp V Mac* *praep cl Misen* *praef leg VI Victricis* *dux legg coh alarum* *Britanicianarum* *proc Liburniae iure gladii*	M. Aur.-Commodus	CIL III 1919=IDRE II 303 CIL III 12224=IDRE II 304	Epetium Salonae
		IIII Flavia			
6	C. Caesius Silvester	*pp IIII Flaviae* *praef castr IIII Flaviae*	Mid 2nd c. AD	CIL XI 5696=IDRE I 126	Tuficum
7	M. Calventius Viator	*curam agens eq sing Augusti*	Trajan-Hadrian	CIL III 7904 AE 1915, 42 CIL VIII 2532=18042	Sarmizegetusa Gerasa Lambaesis
		Other legions			
8	M. Ulpius Caius	*praef XIII Geminae* *praef IIII Flaviae*	AD 213-215	CIL III 1178 = IDR III/5 436 IDR III/5 448	Apulum Apulum

Table II.5. Post-centurionate careers of centurions from Dacia.

II.2.4. Particularities of the legionary centurionate in Dacia

As far as origin and access to the centurionate are concerned, Dacia cannot stand out from the tendencies noticeable for the whole of the Empire in the 2nd-3rd centuries AD. Fortunately, the Roman rule in Dacia overlapped a period of multiple changes in the Roman military system: from the original organization of the Principate in the 1st century to the mobile field army from the middle of the 3rd century. Also, the period of the most radical changes in Roman military organization, that of the Severan dynasty, can be detected in Dacia and in the ranks of its centurions.

Most centurions attested in Dacia do not mention their origin; in fact the vast majority of centurions in Dacia are but simple names, written on stone, next to the general title of *centurio legionis*. Still, in the case of 47 of them – meaning a little more than a third – origin was established, sometimes approximately, other times very exact. About a third of them, meaning 34%, are of Italian origin. This percentage is not very large, given the fact that it truly represents 12,7% of the total number of attested centurions. It is in fact almost equal to that of centurions coming from the European side of the Empire, meaning the Western and Illyrian provinces, without Greece and Africa, 30%. The percentage of those coming from the East is a little below that, meaning 27,5%. The remaining 8,5% is made up of the four centurions originating in Africa but who can be easily included in those coming from the West, in which case that percentage would go up to almost 40%, the largest one. It is obvious that so many Roman names can hide just as many origins and for about half of Roman Dacia's history this might present some relevance, but the situation cannot be very different from the rest of the Empire. To this point it is self-understood that a first generation of centurions from Dacia came along with the legions camped here, being of different origins, Italians but also provincials. At least 30 years, if not 40, had to pass before centurions originating from Dacia were allowed recruitment in the legions garrisoned here. In fact this is not usual and also unlikely because local recruitment would be reduced to auxiliary troops, legions and then, after military practice had become a habit, could we speak of a resource of centurions. When taking these facts into consideration it is likely that we can speak of a local practice of centurion recruitment only from the Severan epoch onwards, when the military environment tends to bind itself to its location and to allow for more and more consistent exchanges with the civilian one.

Besides the centurions who mention their origin as Italy, there are others who give rather exact details about the province they come from: three from Moesia Superior, two each from Dalmatia and Numidia, one each from Moesia Inferior, Noricum, Pannonia Superior, Asia, Africa *proconsularis*, Lycia, Phrygia, Cilicia, Syria, Spain or Macedonia. We can thus notice that they originate both from neighboring provinces and from Greek-Eastern environments.

Regarding their access to the centurionate, should we divide the 14 centurions who offer any detail about their pre-centurion careers on the three recruitment pools for legionary centurions, we notice a clear domination of the *ex caliga* recruitment pool, with exactly 65%, meaning nine attestations. Again we must keep in mind the vast majority of centurions who do not offer such details, meaning almost 90% of those attested, can be promoted from among legionnaries. A number of three knights reached the centurionate and also served in the army from Dacia, two of them came from praetorian soldiers, from which one became centurion and *trecenarius*, another came as a former praetorian centurion and one each from the praetorian fleets or *castra peregrina*.

Besides the origin and access to the centurionate there is also the problem of the chronology of the careers. Concerning this we must focus on the careers that contain any chronological information. The factors considered are the age of recruitments, the length of service prior to becoming centurion, for how long he served as centurions and possibly the length of subsequent service. The fact that there are very few chronological attestations in the careers of the centurions from Dacia makes such a research impossible. From the five such attestations we can determine the length of service prior to becoming centurion in just two cases – 20, respectively 18 years; in four cases we know the length of the centurionate: 20, 14,19 respectively 7(?) years and in three cases we know the *stipendia continua* – 40, 45, respectively 25 years.

The attestations regarding the centurion's posts are just as few in number: only 16. In the present case the absolute majority of attestations belong to *hastati*: one *X hastatus posterior*, one *VIIII hastatus prior,* one *III hastatus posterior*, two *I hastatus posterior*, two collective attestations of the *hastati* from the cohorts *VIII* and *V* and no less than four *I hastati priores* (simply *hastati*). The other are divided in: one *II princeps prior*, two *I principes priores* (simply *princeps*), one *I princeps posterior* and the collective attestation of the *centuria IIII pili posterioris*. The preference that *hastati* have for mentioning their posts is interesting. No theory concerning the centurion hierarchy considers them important, outside of the *I hastatus prior*, but still they seem to dominate this field of attestations and also in Dacia. This confirms that the hierarchy of the centurions from the cohorts II-X is not as rigorous as believed and probably seniority and thus the experience of the centurion was his true rank. It can also be presumed that the young recruits were integrated in the *centuriae* of *hastati* thus they had to be led by centurions with both experience and knowledge in order to train them. From a different point of view, besides the *primi pili*, nine *primi ordines* are attested in Dacia, more than half of the total number of such attestations.

To this we could add the multiple attestations of centurions, by cohort, from the centurion lists of Potaissa.[773] However, outside of the ones in the 1st cohort,

[773] ICLP 10-11.

where order and hierarchy are obvius, no other such categorization can be done, that is why they must be excluded from this statistic.

The most obvious manifestation of the centurionate career is the transfer. This is observed in a number of inscriptions from Dacia, but in smaller numbers than we would have hoped for. Out of 122 attested centurionate careers, a vast majority, 102, meaning 81%, mention only a single centurionate or, better said, the development of a career in a single legion. Following that, the percentage drops dramatically: attestations of service in two legions – 3, in three legions – 6, in four legions – 1 (one legion and the cohorts in Rome - 1), in five legions – 5, in six, seven, eight and ten (11 centurionates) legions – 1. This proves again that, in most cases, a centurion served in one legion, possibly even the one in which he had enlisted. Transfer is an exception not in its absolute meaning but as a means to solve a need for command in one legion or another. The divided statistic, per each of the two Dacian legions, looks as it follows: *XIII Gemina* – from a total of 45, 32 (71%) have only one centurionate; *V Macedonica* – from a total of 62 (59 if three names are identical), 58 (94%) have only one centurionate.

There are a number of 20 attested careers of centurions that mention transfers between legions. Centurions from *XIII Gemina* mention most of them. Five centurions have their first posting as centurions in *XIII Gemina*. One of them is dated in the period immediately following the conquest of Dacia, being immediately transferred in *I Adiutrix*, which probably had a vexillation in Apulum.[774] Another one, who serves during the time of the emperors Elagabalus and Severus Alexander, comes in Dacia for his first posting as centurion, for the next three centurion posts goes in the East, returns for the last two in Hispania and Africa and finally becomes *primus pilus* in *XX Valeria victrix*.[775] A knight who originates in Africa serves his first centurionate in Dacia, for the second returns to Lambaesis and dies during a campaign, while he was serving his third term as centurion in *II Augusta* in Britannia.[776] An unknown character is transferred from *XIII Gemina* to *XV Apollinaris*, probably with some vexillation; he ends with a centurionate in *XXX Ulpia victrix*.[777] Finally, a Spaniard is made centurion in *XIII Gemina*, then passes in three stages through both *Pannoniae* and Britain, finally returning home in *VII Gemina*, where he also holds the primipilate.[778]

Three centurions serve their second centurionate in *XIII Gemina*. A knight from Italy who, following two equestrian militias is decorated by Trajan following the Parthian expedition and becomes centurion in *XXII Primigenia* then in *XIII Gemina*, probably among the *primi ordines*, then becomes *primipilus* in *III Augusta* and has a successful equestrian career. In this case we are certainly dealing with transfers between legions that had vacant posts among the *primi ordines*, in order to become as quickly as possible *primuspilus*.[779] A former who apparently *aquilifer* moved along the whole Rhine-Danubian frontier, was serving as centurion in *XIIII Gemina, XIII Gemina* and *XXII Primigenia*, probably during the wars for the conquest of Dacia, when three of these legions are nearby.[780] Another unknown soldier comes from *XI Claudia* and then goes in the East, in *IIII Scythica*.[781] But the one holding the record for centurionate postings connected with Dacia – 11 – seems to pass through the province serving in *IIII Flavia* or *XIII Gemina*.[782] The centurionate in *XIII Gemina* is the last one in only two cases, both of them seemingly transfers from the East.[783]

During the time that *V Macedonica* was stationed in Dacia we have only five careers with transfer connected with it. In the first case it is a *trecenarius* who probably needed a post among the *primi ordines*, which is why he came to Potaissa, afterwards probably going as *primipilus legionis I Adiutrix*.[784] The second case we are dealing with six transfers: the third and the fourth take place in the legions from Dacia: *XIII Gemina* and then *V Macedonica*; following this he moves to Carnuntum.[785] In any case all transfers take place between the two Pannonian provinces and Dacia. Another possible knight begins his career as centurion in two Eastern legions, then comes to the Danube and serves his last centurionate and his primpilate in *V Macedonica*.[786] Following this, he will have a successful equestrian career. There is also one coming from the East who serves his first centurionate in *V Macedonica*, then leaves for the Orient and finally returns on the Rhine.[787] Finally, a former praetorian centurion seems to become the legions' *I princeps posterior*, probably awaiting for the primipilate.[788]

Three such careers are attested in the *IIII Flavia*. A former pretorian decorated in the Dacian wars becomes a centurion in the *II Augusta*, in Britannia, then arrives in Dacia, in the *IIII Flavia*, serves on two centurionate posts in the Orient and returns on the Lower Rhine. Finally he will become *primipilus* and *praefectus castrorum* of the legion *IIII Flavia*.[789] Probably a former legionary cavalryman and centurion from *IIII Flavia*, after he was Nigrinus' *exercitator equitum singularium* when he governed Dacia,[790] was transferred in the *V Macedonica*, but at Troesmis, leaving with Hadrian for the East and becoming *curam agens* of his cavalrymen guards. Finally, an Italian knight was a centurion in five legions,

[774] C2. Piso 2000, 205-206, *contra* Opreanu 1998.
[775] C10. Petolescu IDRE II 431.
[776] C11. Petolescu IDRE II 443.
[777] C24.
[778] C45.

[779] C8.
[780] C14.
[781] C23.
[782] C13.
[783] C3 and C9.
[784] C46. Petolescu, IDRE I 133.
[785] C52.
[786] C53.
[787] C54.
[788] C71.
[789] C108.
[790] C109. Piso 1993, 19-23 and Petolescu 2005, 195.

probably in connection to the Dacian wars, remaining in the new province with *IIII Flavia*.[791]

Thus, we can notice that the career paths of those who passed through Dacia are various. They come in Dacia from the following provinces: Germania Inferior and Superior, Pannonia Superior and Inferior, Moesia Superior, Britannia and the East, two come from Rome. They leave for: Africa, East, Germania Superior, Moesia Superior and Pannonia Inferior. The number of transfers is not large enough that we could reach any conclusions, but it can be stated that *XIII Gemina* has a preference in receiving centurions from Germania Superior (3) and the Eastern provinces (3) and to send them in the same Oriental provinces (3) and Africa (3). In the case of the other two legions we cannot make such remarks; we can only state that *V Macedonica* receives two centurions from Pannonia Inferior.

As far as the transfers inside the same provincial army, few things can be said. If only one such example is to be considered, with other few possibilities since the time of the creation of the province, we have yet another argument to the singularity and the statics of the centurion career. If transfer came very rarely, even at as close as 40 miles away, then it cannot be the custom.

Finally, as far as future careers are concerned, from the six attested examples in Dacia, two stop at the primpilate, two become *praefectus castrorum* and two are equestrian careers that also include procuratorships.

[791] C110.

II.2.5. The role of centurions from Dacia in the provincial society

We have already discussed the economic role that centurions play in the Roman world. Dacia does not stand out because of attestations of such situations, where centurions are seen involving themselves in the economic life. For example, even if we have no attestations of centurions involving themselves in the mining works from Alburnus Maior, they must have been present here, just as they are also present in other mining areas of the Empire, either to keep order and ensure the proper functioning of the enterprises or for technical expertise or leading the worksites. In any case there is other evidence of soldiers in the area; there is a legionary who is a witness in the contract of selling-buying of a slave girl.[792] Also the newest discoveries of archaeological materials and tile stamps of the legion *XIII Gemina* might indicate that the central administration was present here, possibly through its officers. Otherwise centurions, along with their soldiers, have clearly been involved in setting up the infrastructure of the province. Concerning this we have several examples. At least three centurions are attested leading vexillations involved in constructions. Two of them command the legionary vexillations that have built the auxiliary fort of Horghiz or have rebuilt the *tabularium* of Apulum. The third seems to command the entire *numerus Palmyrenorum sagittariorum* mobilized for rebuilding the temple of Bel from Porolissum. The fortification of Romula at the half of the 3rd century AD has already been mentioned. In this context we find mentioned at least six legionary *centuriae* from the legions *VII Claudia* and *XXII Primigenia*, even from Germania Superior, for this construction work. Last but not least, the centurion from the legion *IIII Flavia* from Drobeta, who seems to be a *regionarius*, must have been sent to command a *regio* under military control. We cannot know where such a *regio* was placed, although the location would be of great interest. A similar case could be that of the centurion of *X Fretensis*, dedicating an altar to Iupiter *optimus maximus* on the northern frontier.[793] The succession of letters after his name and title is *RE/G*, which could in fact point out to a *centurio regionarius*. His role and purpose in this remote, border region is difficult to find, as well as the dating. However, he is a centurion in charge with an *ad-hoc* military region.[794]

Centurions do not seem to leave their mark in Dacia's social life, at least actively. Three of them are part of the equestrian order but none of them originates in Dacia. No centurion holds any municipal office in the cities of Dacia; only one enjoys the status of *bouleutes* in Nikopolis *ad Istrum*. Another centurion is the son-in-law of a decurion from Sarmizegetusa but this says nothing about his social status; another is the client of a procurator. Thus, from this point of view, we cannot say

[792] IDR I TabCer VIII.

[793] CIL III 7625, Domneşti, near Bistriţa. Dabrowa 1993, 17, note no. 51, acknowledges the inscription, but not the centurion, as he doesn't appear in the prosopographic lists. He just argues that this is no proof of any vexillation of *X Fretensis* in Dacia Porolissensis.

[794] Details in Piso, Cupcea 2014, forthcoming.

that centurions involve themselves in provincial society – this would also be very difficult, if not even impossible first of all because they are attested while still serving in the military and second because many of them are foreigners. Add to this that a centurion's life can be very mobile, thus any kind of social-cultural allegiance to a community, different than that of origin, is purely incidental.

Judging from all points of view we see the centurion perfectly integrated in his legion but also in the provincial army as a whole. In Dacia there are several commemorations or praises of this *esprit de corps*: at least one case can be attributed to centurions. We are referring to the dedication set up by the centurions, one from *V Macedonica* and the other from *VII Claudia commanuculis*.[795] The relation of comradery is also strengthened by both centurions originating in different provincial armies but with this '*esprit de corps*' specific for the Roman army, especially in the 3rd century. Another such display is represented on the building plaque of the *schola speculatorum* from Apulum,[796] where we find reunited as members of this 'professional association' those holding the post of *speculator*, former *speculators* who became *commentarienses*, *cornicularii* or centurions, in both legions from Dacia,[797] all members of the *officium consularis*; such a case could just as well be also the construction plate of the *schola signiferum* in Potaissa, were it preserved better.[798] This proves how the central provincial administration is a very important factor for the cohesion of the military forces from the province especially because they recruit people from as many units as possible.

We cannot offer a great deal of information concerning the intellectual preoccupations of the centurions from Dacia because we lack proof. Still, the example of the funerary monument of one centurion from *IIII Flavia*,[799] which is partially in verses, along with a stylization of a transverse crest shows that centurions are accustomed to literature, education being essential in order to obtain their rank. Finally, 'name tags' on personal items, as the piece of armour and bronze plate from the fortress of Potaissa can be another clue on the literacy of the soldiers and centurions.[800]

Another way to integrate centurions in the mass of the soldiers is religion. Although his religious role drops dramatically in the legion as he is overtaken by the *primus pilus*, the centurion has the mission to mark the fulfillment of the missions vexillations are sent on or to officialize in any way any religious act of the *centuria*. In the case of Dacia nearly 60% of the votive monuments set up by centurions have been discovered in fortifications. The same percentage is true for the dedications for the supreme deity, Iupiter, in all its forms.[801] When analyzing its epithets, besides *optimus maximus*, *victor et depulsor*[802] stands out. Also, several legionary centurions from Sarmizegetusa and Potaissa set up altars in honor of Iupiter *optimus maximus* Heliopolitanus,[803] along with a centurion from *XIII Gemina*, in Syria;[804] these acts are probably connected with them taking part in Trajan's Parthian campaign. Dedications for several deities include those for the Capitoline triad[805] and those for Iupiter *optimus maximus et dis hospitibus huiusce loci*.[806] At Potaissa altars for Mercurius Augustus or Hercules *invictus* are also set up. A particular religious manifestation from Potaissa is attested by several altars dedicated to *genius centuriae*,[807] very rarely set up by the centurion.[808] Healing gods are also present in the dedications set up by centurions, especially the nymphs from Germisara and Aesculapius and Hygia. Several dedications are set up also for Apollo, even a special one for *Phoebus* Apollo *Parthicus*;[809] its presence in this context is connected to Lucius Verus' Parthian campaign. A centurion sets up at Potaissa a monument for the sacred Egyptian pair, Isis and Serapis.[810] Other gods worshipped by centurions are Iupiter *optimus maximus* Dolichenus, *Deus Aeternus*, Diana, Silvanus or *Eponae et Campestres*. Because they command the *centuriae,* centurions are also representatives of legionaries in their relation with the divinities but most of their religious acts are performed in their own name and in the military environment.[811] We cannot assume certain religious preferences for centurions, least of all of those from Dacia, who share the pantheon with the other soldiers.

Although Dacia does not stand out when compared to other provinces from the Empire when it comes to the careers of the centurions, still it offer substantial information for the study of this phenomena in the whole Roman world. By studying only the centurions from Dacia we will never reach a plausible conclusion but by integrating them in the universal system of management and organization of military resources, we can find points of real interest.

The centurions from Dacia are the most active part of the Roman soldiers found here. They have both the authority and the possibilities to stand out while in service. Still, most of them are simple names and indications to what legion they were serving in. This fact can lead to a less bright conclusion than one would have expected: a

---

795 AE 1976, 574 = ILD 499, Potaissa. For the usage of the terms, see Lendon 2006.

796 CIL III 7741=14479 = AE 1901 154 = IDR III/5 426.

797 For a detailed discussion see Cupcea 2008.

798 CIL III 922, 927, 928, 929 = ICLP 6, Potaissa.

799 CIL III 1653=8143 = IDR III/1 1, Kovin, Serbia.

800 C106 and C107.

801 Popescu 2004, 277.

802 AE 1944, 28 = IDR III/5 232, Apulum.

803 AE 1933, 18 = IDR III/2 243, CIL III 1353 = IDR III/3 95, CIL III 1354 = IDR III/3 96.

804 AE 1928, 84 = IDRE II 409, Nasranîyé (Beyrouth), Syria.

805 CIL III 7764 = IDR III/5 197, Apulum.

806 AE 1992, 1470 = ILD 464, Potaissa.

807 CIL III 7672, ILD 501, 502. Bărbulescu 2004, 376.

808 *Contra* Popescu 2004, 278, who is convinced that Aurelius Asclepeiades (ILD 501) is also a centurion. This is highly unlikely because his name is in nominative form, being the dedicant.

809 AE 1972, 454=ILD 482, Potaissa.

810 CIL III 881, Potaissa.

811 Popescu 2004, 280.

centurion's life and career were relatively static as he deals with the same things for a lifetime, probably in the same place. The strongest motivation must have been remuneration, but for certain many desired the primpilate and imagined themselves in that position, very few obtaining it however. Still, we can say it represents the 'engine' for the existence of a mass of about 2000 professional officers, educated and trained by the army in order to be later used in virtually every domain.

Part III. *PRIMIPILI* AND *PRIMIPILARES*

III.1. General facts

III.1.1. The rank of *primus pilus* and entry to it

The title of the officer appears in two distinct versions, both in literary and juridical sources and in inscriptions: *primus pilus* and *primipilus*. There is no doubt that the first version is much more used, especially in 1st and the first half of the 2nd centuries AD, mostly in inscriptions from Italy and Narbonensis. From this point forward we notice the contraction of the term and the increased usage of the form *primipilus* until its exclusive usage in the 3rd and 4th centuries. However, its meaning remains the same during the first four centuries and in all sources.[812]

The discussion concerning the Republican primpilate can start based on a passage from T. Livius,[813] who claims that 23 centurions *qui primos pilos duxerant*, are recalled in the army on lower positions. One of them presents his exceptional career as a centurion and is made *primus pilus* in the first legion (171 BC). Apparently the rule was to recall and send them to serve on different positions in each campaign but we cannot rely too much on these early examples. Another important source is Caesar who confirms the occupation of the post of *primus pilus* for one year,[814] or the possibility of its occupation at least twice.[815] As for the age requirement for serving in this position we know from a passage[816] that the man was the father of an adult so he must have been over 40 years of age. Anyway the most important criterion for obtaining this position is *virtus*.[817] At that time they are most frequently referred to as *centurio primi pili*, and are considered the highest-ranking centurions in a legion. They are part of the commander's *consilium* for one year, while they occupy the position.[818] Polybius is the one who includes the *centurio primi pili ex officio* alongside tribunes in the general's staff.[819] Further promotion options lead towards tribunes but they could no longer be recalled as centurions, such situations being rare, maybe only during the triumvirate.[820] It is possible now to hold this post several times, sometimes even consecutively.[821] Augustus took his adoptive father's habits, appointing *primipili* as commanders of auxiliary troops (thus on positions of prefects or tribunes).[822]

During the Principate, after serving as centurion between 15 to 30 years, after transfers in several legions and after promotion within the *primi ordines* a centurion about the age of 50 could be *primus pilus* for one year. This was the fastest way up the social ladder[823] and the highest rank accessible to the professional soldier.[824] Only the best and the most favored centurions reached the first cohort and many retired after becoming *primus pilus*.[825] Although he is the commanding officer of the first *centuria* from the first cohort he has much greater responsibilities. The post is held for one year, is very well paid and can provide access to the equestrian order.[826] He places immediately below the tribunes as chief-centurion of a legion, holds in custody the eagle (confirmed by the eagle appearing on their representations as a distinctive sign). The ambition and dream of every centurion, or even soldier, was to join the *primi ordines* and to become *primus pilus*.[827]

At the beginning of the Principate *primus pilus* was promoted to *militia equestris*. Centurions who were beneath this rank can be promoted to the *praefectura* of an *ala* or of a cohort, but only *quingenariae,* while in the case of the *primus pilus* this type of promotion is made only to *milliariae* units. These tasks, similar to the detaching of legionary *principales* to command posts in *auxilia*, were meant to strengthen the unity of the army.[828]

In 64% of the *primipili* careers we have no information concerning the ranks held while centurion, before reaching the primpilate. Most of them only mention *centurio legionis*, without further details and this gap prevents us from drawing exact conclusions. Opposite to what many scientists believe – namely that those former praetorians who become legionary centurions and *primipili* have an easier access to procuratorships – there is no evidence to back it up, the *primipili* from the praetorian guard have access also to a more modest career that might culminate with the post of *praefectus castrorum*. As far as centurions who previously served as praetorian petty officers, be they *cornicularius praefecti*

812 Wolff 2004, especially 22.
813 42.32-35.
814 *BG* 5.35, *BC* 3.91.
815 *BG* 1.13.
816 *BG* 5.35.
817 *BG* 5.44. In this context two *primi pili* are mentioned in Caesar's book: T. Balventius, brave and authoritarian (*viro forti et magnae auctoritatis*), with both legs wounded and Q. Lucanius, also a *centurio primi pili*, who dies fighting bravely (*fortissime pugnans*), helping his son. Brown 2004, 300. It is also Caesar (*BG* 2.25) who, when speaking about the battle on the Sambre, says that a *primus pilus*, Sex. Iulius Baculus, a very brave man, exhausted by his numerous wounds, could not stand up any longer. Goldsworthy 2008, 174.
818 Campbell 2001, 326.
819 Keppie 1984, 20.
820 Dobson 1978, 3-4; Birley 1978. One proof is the naming of auxiliary units according to their commanders, *primipili*, since Caesar's time (*ala Scaevae*).
821 T. Livius 42.34.5-11. Spurius Ligustinus, the centurion with an exceptional career serves four times as *primus pilus* in a time span of only a few years. Keppie 1984, 36.
822 Other occasional activities of Republican *primi pili*: commanding the army during Hannibal's invasion (Livius 25.19.9), setting up outlaw lists during Sulla's dictatorship *etc*., irrelevant for their usual jobs. To conclude, the Republican *primus pilus* is the highest-ranking legionary centurion for the course of one year, though he serves a couple of times on this post during his career, and represents the legionaries in the general's staff. He might be below 50 years of age but still around 40. Experience, virtue and bravery in combat are required. Promotion chances are very small; evidence for superior posts such as *praefectus castrorum*, urban tribunates or large sums of money offered at discharge are missing. Dobson 1978, 5.
823 Junkelmann 1986, 112; Le Bohec 1994, 44.
824 Allen 1908, 6.
825 Rankov 2008 ,41.
826 Webster 1969, 118.
827 Keppie 1984, 151; Junkelmann 1986, 112; Campbell 1984, 47.
828 Domaszewski 1908, 112.

or *evocati*, only a small part of them reach the primpilate and from them even fewer obtain procuratorships or higher prefectures.[829] It is certain that there were many legionary centurion positions held by former praetorians, especially in the *primi ordines* but this can be explained away according to circumstances, not necessarily as an intention pursued by emperors who ruled before Severus to maintain the Italian characteristic of the centurionate. In the pre-Flavian period it is apparent that *primipili* who served as legionaries were preferred for occupying the posts of tribunes of the troops from Rome. It is likely that the experience of AD 69 required a greater unity of the two kinds of units, opening the door for praetorians to be promoted among legionary centurions.[830]

From another point of view it is noticeable that the chances of a soldier to become *primus pilus* or even higher were almost nonexistent. Most *primi pili* that go far begin their careers with the primpilate or legionary centurionate.[831] There is also the possibility that those who start their careers as centurions skipped the lower posts because of their obvious loss of importance. The ones who were promoted to *primus pilus* and beyond were those who began their careers as centurions because they belonged to an upper class.[832]

According to calculations 20 to 30 years were needed to reach the primpilate.[833] The *origines* coincide with those of centurions. At first the recruitment pool was confined to Italy and the colonies in its nearby vicinity, then, gradually, it extends to include well-Romanized provinces up until the period Trajan-Hadrian when the provincials and Italians are in equal numbers. During the reigns of Antoninus Pius and Commodus Italians have again priority but after AD 193 their numbers dwindle and even disappear from the ranks of the *primi pili*. During this whole period they originate from the ranks of the legionaries, from praetorians and from directly appointed centurions. While during the period Antoninus Pius-Commodus Italians have priority – and thus also officers from the guard – the situation changes with Severus, *castra peregrinorum* and provincials, even auxiliaries, being granted priority.[834]

Direct access to the primpilate is usual only in the case of knights.[835] Among them there are those who reach this from posts reserved for *primi pili* or who are directly promoted *primipilaris* thus they do not even serve in the legions as centurions or *primus pilus*. It is peculiar that in their subsequent careers they will not hold any exceptional positions, with few exceptions. For a young knight or municipal magistrate a career on a post of legionary centurion and a primpilate was very profitable. While a regular centurion reached the primpilate around the age of 50 those appointed directly had the time, prestige, education and greater chances to become *primi pili* faster. Plus they obtained a profitable job for life. The regular equestrian career required money and offered no certitude after finishing it. The centurionate and primpilate offered a possibility for an exceptional social and economic status, as proven by their eventual careers. Few obtain tribune positions in Rome and only one a procuratorship. This change in direction came at around the age of 30, though examples are also found at the age of 18, 19 years.[836]

There is another – very profitable – way to reach the primpilate besides the legion and praetorian cohorts, developing only at the end of the 2nd and beginning of the 3rd century AD. It involves gaining access through the *castra peregrina*: from the post of *frumentarius legionis*, from *centurio legionis-centurio frumentarius*, or directly from *centurio frumentarius* because of the great importance placed by Severus and Caracalla on these units. Some can be *navarchi* of the fleets or praetorian officers before being transferred in the *castra peregrinorum*.[837]

To conclude, those who reach the primpilate can be divided in three groups:

A: legionary soldiers: legionary petty officers who could reach the centurionate only after 13-20 years of service until the reign of Severus.
B: praetorian soldiers: urban or praetorian petty-officers who can obtain this position only after serving as legionary centurion, as legionary and praetorian centurions or only as praetorian centurions (16 years).
C: centurions appointed directly: centurions who promoted directly, only through legionary centurionate or both categories.[838]

The continuation of a career above the ranks or *primipilus* and *praefectus castrorum* was an exception to the rule. Those who offer details about their careers prior to the primpilate usually end their career here.[839]

[829] Birley 1941, 200. In fact statistics would look like this: out of 29 centurions who became *primipili* 11 have occupied posts after the primpilate and only two enjoy a career as *procurator*. Out of 36 praetorian soldiers who became legionary centurions only 15 reached the primpilate and six were further promoted. These numbers do not indicate that the chances of the praetorian to promotions beyond *primipilus* are higher than those of any other type of legionary centurion. The references are based on a list of 101 *primipili*; 34 of them indicate their inferior carer and 20 (22) originate in legions. Of course, looking at the numbers it seems that praetorians are favored for promotions.

[830] Birley 1941, 199, 202-203. An unique situation is that of L. Maximius Gaetulicus, who in AD 184 fulfilled a promise as *primipilus legionis I Italicae*, at Novae, which he made 57 years ago as *tiro aput legionis XX Valeriae victricis*. It is a unique case, not all young recruits could hope for this, but it confirms the hope to become *primus pilus* and the fact that is was the main stimulus for those seeking centurionate and wishing to serve much longer than was normal.

[831] CIL VIII 9045.

[832] Allen 1908, 9-10. Out of the 72 examples of *principales* becoming centurions only 10 seem to as far as the primpilate. Praetorians seem to have an advantage over legionaries because of their shorter term of service and the possibility to become *evocatus* faster.

[833] Dobson 1978, 16.

[834] Dobson 1978, 38-42.

[835] Frontinus, *Strat.* 4.6.4; Plinius, *Ep.* 6.25; *HA, Pert.* 1.5.6; Dio 52.25.6-7.

[836] Dobson 1978, 44-48.

[837] Dobson 1978, 49-50.

[838] Dobson 1978, 50-52.

[839] Dobson 1978, 59.

Another problem that is raised is the age at which one can become *primus pilus*: Blossius Pudens dies at the age of 49, immediately after being named *primus pilus* or after serving one year. Apparently 50 is the right age for the primipilate though there is evidence for even more advanced ages: Juvenal,[840] Pliny.[841] There is epigraphic evidence for holding the primipilate when one is very old: 72 years, more than 69, 45 years of service before the prefecture of the legion, 58 years of service; this is not a later habit as proven by Suetonius.[842] The centurion's length of service varied between 15 and over 30 years and there is no difference between legionary centurions and those from Rome from this point of view. These numbers fit well with the age for the primpilate: between 50 and 78 years of age.[843]

The position of *primus pilus* was not available for everyone regardless of ambitions. A good education and useful administrative skills were at least as – or even more – desirable than physical strength and military skill. Those who possessed these qualities had the chance to join the military on high ranks, in the praetorian cohorts or directly appointed centurions.[844]

[840] 14.197: *ut locupletem aquilam tibi sexagesimus annus adferat.*

[841] *Nat. hist.* 14.9: *quid quod inserta castus summa rerum imperiumque continet centurionum in manu vitis et opimo praemio tardos ordines ad lentas perducit aquilas.*

[842] *Calig.* 44: *Plerisque centurionum maturis iam et nonnullis ante paucissimos quam consummaturi essent dies primos pilos ademit.*

[843] Dobson 1978, 60-63.

[844] Watson 1969, 88.

III.1.2. The career of the *primipilares*

All centurions who held the position of *primus pilus* form a distinct class or order, *primipilares*,[845] and enjoy special payments and benefits. They are divided in two classes: those who retire from the army and those who follow a career of officers and eventually procurators/prefects. During the Republic, the borderline between simple soldiers and knights could not be crossed. During the Principate the primpilate was the gateway to the equestrian order. Since the census requirements still applied, the emperor offered them a large sum of money after discharge, meaning after the primpilate.[846]

At the beginning of the Principate the command of auxiliary units was tasked to senior legionary centurions, especially *primi pili,* a custom adopted by Augustus from Caesar. This custom was meant to keep auxiliary units updated with the Roman-Italian military practice. Now even auxiliary centurions and decurions are recruited from among legionaries. Still, *primi pili* shared the command of these troops with prefects who had a barbarian origin, tribal chiefs.[847] The solution for compromise was that it was possible for an *ala* to be commanded by a *primus pilus*, by a legionary centurion or by a local noble as well as by equestrian officers, previously trained.[848]

Augustus is the one who, by means of experiment, will recognize the value of a very experienced corps of officers. His first innovation is the title of *primipilaris*, the former *primus pilus*, who from this point forward cannot return to the centurionate but has other options presented to him. Also it is Augustus the one who establishes the large sum of money to be paid upon discharge as well as their banding together in the *numerus primipilarium* in order to supplement strength and military vigor where needed. It is probable that the primipilate was still occupied for one year according to the Republican custom. The post of *praefectus castrorum* was also invented now[849] although an example exists as early as Caesar. Anyway this *praefectus castrorum* does not mention his legion so he has no clear affiliation, up until Claudius.[850] Augustus maintains two promotions characteristic for this period: towards *praefectus fabrum*[851] and *praefectus classis* (probably instead of posts of procurators that did not exist yet), needed to be filled with experienced soldiers. Tiberius maintained the system[852] and Caligula did not have time for innovations. The experience of the occupant and the stringent needs

[845] Definitions and general notions in *RE* (Lammert 1954) and *DNP* (Campbell 2001).

[846] Allen 1908, 6.

[847] Cheesman 1914, 91.

[848] Birley 1978, 259, 262.

[849] The first mentioning dated under the Principate is in AD 11. Keppie 1984, 149.

[850] Dobson 1978, 6-7. According to R. Syme it seems that there was one for every legion. Syme 1932, *passim.*

[851] A very important post in the governor's staff. Dobson 1966 and Domaszewski-Dobson 1967, xxix.

[852] Still the title of *primus pilus II* seems to date from his period (ILS 2687). Parker 1928, 198.

were selection criteria but still, some careers start to shape up. At about this time appears the post of *tribunus cohohrtis praetoriae*, an independent posting, and *praefectus cohortis (equitatae)*, the two types of careers of *primipilares* starting to separate.[853]

In these pre-Claudian times, *primus pilus* was regularly promoted to the tribunate of the legion immediately after holding the prefecture of an *ala* or cohort. Also now legionary centurions and the *primi pili* receive administrative tasks in a *civitas*, except that the *primi pili* become *praefecti levis armaturae* in the Alpine provinces, Illyricum or Corsica or *praefecti civitatium*, meaning the leader of the civil administration.[854]

Emperor Claudius is the one who sets military careers on a new direction. With exceptional strictness he manages to create an equestrian career based on the fulfilling of *militiae*.[855] Inside this system the new knights created by Augustus, the *primpilares*, will have their own distinct and well-defined place, with multiple opportunities to serve the emperor after fulfilling their professional military service. By the time Britain was conquered the rules were already made, with regards to the direct promotion of the *primus pilus* that finished his one-year service to *praefectus castrorum* and thus an attempt at organizing the equestrian career was made. The post of *tribunus militum* disappears from the careers of the *primi pili* because there is a need for experienced officers in the cohorts in Rome. Also *praefectus equitum* disappears and is replaced by *tribunus cohortis praetoriae*, obtainable immediately after the primpilate until the reign of Nero when the other two urban tribunates are introduced. Under Claudius, or perhaps earlier, the title *primipilus II* appears with its new meaning; it is held after the tribunate of the praetorian cohort, it can lead directly to a procuratorship and altogether the old meaning of *primipilus bis* disappears.[856]

Claudius however opens two possibilities for *primipilares,* allowing them to develop their career in both directions. The most prestigious way for progressing hugely on the social scale is to start an equestrian career. The emperor reserved a special echelon of equestrian *militiae* and thus opened access to more military oriented procuratorships or prefectures, in principle different from the ones reserved for knights. The second option reserved for *primipilares* is to end their careers while holding a very important rank in the legion, namely that of *praefectus castrorum*, also established now and which becomes the third most important post in the legionary upper hierarchy. Claudius created the post of *primipilus II* for those choosing an equestrian career in order to maintain active professionals who were about to be sent to occupy procuratorships or prefectures.[857]

Besides having the command of the first *centuria*, the *primus pilus* was a counselor of the legate. He probably spent more time in the general's staff in order to become a strategist officer for the whole legion. He has a special relation with the eagle, sets up monuments in its honor and wears its symbol.[858] By comparison with a *princeps* he has no administrative duties because he does not need them. His post is prestigious, the highest honor available to a centurion, for one year he is *primus inter pares*, the highest-ranking and most honored of the centurions, privileged and for the rest of his life a *primipilaris.* Evidence of his privileged status is also the fact that *praefecti castrorum* continue to refer to themselves *primipilaris*, while tribunes or *primipili II* prefer these other titles.[859]

Many *primipilares* withdraw immediately after the primpilate but the emperor does not exclude them from his future plans.[860] Those who did not withdraw were gathered in Rome in the *numerus primipilarium*, from where they can be sent to a commander's staff. This is a veritable reserve of professionals needed for different jobs in the imperial administration.[861]

The body of *primipilares* served in Rome, in a *numerus primipilarium* attested by Hyginus,[862] living in the fortification with the *evocati*, close to the praetorians.[863]

[853] Dobson 1978, 7-9; Dobson 1979, 199. *Praefectus levis armaturae* also appears, probably a command rather than a post, *praefectus civitatium*, a post involving both military and civilian duties.

[854] Domaszewski 1908, 113. The situation of the careers of *primipilares* before Claudius also in Zwicky 1944, 52-53; Devijver 1970, 74.

[855] For a detailed discussion concerning Claudius' reform of the equestrian *militiae* see Devijver 1970, especially 74-75 and Devijver 1995, 179-180, about the clear distinction between the careers of *laticlavi, angusticlavi* and *primipilares.*

[856] Dobson 1978, 11-13.

[857] Dobson 1978, 14. Some small changes take place under Nero then nothing major until Gallienus, not even the *castra peregrinorum* does not influence this order.

[858] Juvenal 14, 197; Plinius, *Nat. hist.* 14, 9.

[859] Dobson 1978, 65. Tacitus prefers the term *vir militaris* for *primipili.*

[860] Suetonius, *Nero* 9.

[861] There is a multitude of such examples in the year of the four emperors (AD 69). They were sent away on emergencies and delicate missions and under Caesar and Augustus need and urgency determined the transfer, not the career. In Rome they had a multitude of tasks: in civil wars they command campaigns, in peacetime they command units from Rome, as curators of the cohorts of *vigiles* or *curator statorum.* Sometimes they have more unusual tasks, such as supervising a canal (Suetonius, *Calig.* 21). They take part in the command of vexillations, sometimes a *primipilaris* can take command himself, in a crisis situation, like Aquilis did during Civilis' rebellion (Tacitus, *Hist.* 4.15), of what remains of the cohorts whose commanders followed him. Another, being in Corbulo's staff commanded two *alae* and three cohorts and in another case a *primipilaris* commanded *equites singulares* in Rome. Besides these ranks they can also hold civilian posts: *dilectator* in Transpadana, commander of the forces fighting against bandits in Italy, retrieving the body of a general in Rome for burial, ensuring the frontier, repairing the roads. They are a connection between the emperor and the local administration from their post as *praefectus Augusti, curator viarum et pontium Umbriae et Piceni.* They are found commanding legionary vexillations especially from the Severan dynasty onwards because senators are gradually excluded from such commands. After the equestrian *militiae* disappear they serve as experienced officers on posts such as *praepositus equitum singularium Augusti* or *praepositus classis Misenatis.* They will replace legates as *praepositi legionis*, then as *duces legionis.* In the civil administration they are used as emissaries, ambassadors *etc.* They are valuable cadre, trained, useful in every situation, used by the emperor in the most delicate tasks. Domaszewski 1908, 116-117; Dobson 1978, 66-67; Junkelmann 1986, 112; Goldsworthy 2008, 72.

[862] *De munit. castr.* 6

[863] Domaszewski 1908, 116.

In Tacitus' writings they are mentioned several times under Claudius and Nero and in AD 69 they are given command as *praepositi* or *curatores*.[864] Following this waiting period they could look towards tribunates in Rome or procuratorships. In the 2nd-3rd centuries AD they command vexillations, *equites singulares Augusti*, praetorian fleets, Egyptian legions, they can even become prefects of the *praetorium*. They have no political role except in times of crisis; these exceptional professionals return to the emperor's *praetorium* in order to serve him in the most varied tasks.[865]

The legionary career. *Praefectus castrorum* and *praefectus legionis*

Immediately after serving their primpilate the officers – who have finished now their military education – could be promoted to *praefectus castrorum* (which indicates the rank of the post, in between the two legionary tribunates[866]). This was the other, easier option available for *primpilares* who could not or would not go to Rome.[867]

The post of *praefectus castrorum* is invented by Augustus, for each legion, but the mentioning of the legion starts only with the reign of Claudius. In the first years of the Principate there are several *praefecti castrorum imperatori Caesaris Augusti* or *Tiberii Caesaris*, formulas that might indicate operations where they served under the emperor's command, as part of his staff. Under Claudius the post of *tribunus militum* is eliminated from the career of the *primipilares* and *praefectus castrorum* occurred, on a lower position in the hierarchy as the post that could be occupied immediately by a *primiuspilus* who had finished his service. In literature however we encounter the title without the mentioning of the legion.[868]

As a new rank, created under Augustus the *praefectus castrorum* was very important and could sometimes command vexillations or be part of the general's staff. Most of them seem to be of Italian origin. Under the first two emperors they hold a variety of posts: prefect of an auxiliary unit or fleet, tribune. They are especially important in Egypt because of the absence of senatorial legates.[869]

Starting with Claudius the legion is also mentioned and in case there are several legions camped in a fortification it is likely that the senior one effectively was in command. *Praefectus legionis* appears only at the beginning of the 2nd century as an abbreviation from *praefectus castrorum legionis*. The last such example dates back to AD 201, without assuming that Severus abolished the position but that it had already transformed a long time ago. We can place the post to its proper place in the hierarchy because of a collective dedication made by tribunes.[870] It is placed after the *tribunus laticlavius* so he would be the third ranking officer in the legion, before the *tribuni angusticlavi*.[871]

After the reign of Claudius, the meaning of the rank is local command of a legionary fortress. It is the end of the career for a *primus pilus* who does not follow the equestrian *militiae*[872] and the highest post obtainable at the end of a professional military career. After this one, sometimes also held for three years, the career of the holder was over but not as a failure, on the contrary. In the 3rd century we discover the repetition of the post *praefectus castrorum* possibly indicating their ever-increasing importance. The last *praefectus castrorum* is dated under Elagabalus or Severus. His duties are not exclusively administrative, he is dispatched with the legion everywhere, and sometimes he commands construction detachments. His experience is used in command in critical situation – such as the Civil War – by offering him command of the vexillation.[873] After Domitian solved the problems with the legionary garrisons, *praefectus castrorum* also mentions his legion. He was a commander of the fortification, responsible with its placement, construction and internal planning.[874] In wartime he organizes the marching, ammunition and siege operation and in peacetime he commands the body of specialists (*mensores, medici, horologiarii*[875]). If the legate and the senator tribune are missing he can take command of the legion.[876] He is appointed by the emperor and under the command of the governor. He has a double role: civilian – supervising senator commanders for the emperor's benefit, military – his experience makes him the actual commander of the legion.[877]

Egypt has a special situation. The exceptions that are present here can help us clear up to a certain point the duties and status of the post of *praefectus castrorum* for the rest of the Empire as well as the transition to *praefectus legionis* and its meaning. In any case, the rank seems to be more important here than in the rest of the Empire.[878] It is possible that these commanders – *praefectus legionis* to have disappeared if two legions were camped together and to be replaced by a *praefectus castrorum Aegypti*.[879]

[864] *Ann.* 2.11, 13.36; *Hist.* 1.31, 2.22, 3.70.
[865] Durry 1968, 21-22.
[866] Campbell 1994, 47; Rankov 2008, 41.
[867] Allen 1908, 8.
[868] Dobson 1978, 68.
[869] Syme 1932, 109-111; Saddington 1996, 244.
[870] CIL VIII 18078.
[871] Dobson 1978, 69. Poenius Postumus (Tacitus, *Ann.* 14.37) is *praefectus castrorum* and emergency commander of the *II Augusta* during Boudica's rebellion, a situation when it is unlikely that all tribunes were absent.
[872] Domaszewski 1908, 120.
[873] Dobson 1978, 70-71, Junkelmann 1986, 113. Sabinius Nepotianius is *praefectus vexillationis legionis I Minerviae*, at Iversheim, from his rank of *praefectus castrorum* of the same legion, sent away with a number of tribunes.
[874] Tacitus, *Ann.* 12.38.3.
[875] Vegetius, *Epitoma* 2.10
[876] Tacitus, *Ann.* 14.37.6; Rankov 2008, 41.
[877] Parker 1928, 191-193; Junkelmann 1986, 112.
[878] The problem was amply debated several times without coming to a definitive solution. Domazewski 1908, 121; Devijver 1974, 459; Dobson 1978, 73; Parker 1928, 193-196; Saddington 1996, 245; Dobson 1982, 322-334.
[879] Dobson 1982, 325-334.

In any case the quantity and quality of superior officers present in Egypt's army is still undecided.[880] Egypt's particular situation, meaning knights who were granted *imperium*, can also include from some point on these *praefecti castrorum*, obviously different from those having the same title from the rest of the Empire. We might expect things to be cleared up by the situation from Mesopotamia from the beginning of the 3rd century but, although similar, things do not seem to be clearer in this case. In the case of *praefectus legionis* we know of two meanings. First, the contraction of *praefectus castrorum legionis*, present from the 2nd century, simply a *primipilaris* on the post discussed above. The second option, *praefectus legionis*, dominates the scene from Severus onwards, him being the one who organized Mesopotamia based on these criteria. Led by a *praefectus Mesopotamiae*, the command of the legions was given to *praefecti legionis I*, respectively *III Parthicae*, commanders of ducenary rank. Even in the *II Parthica*, stationed near Rome, the commander is a *praefectus legionis*, identical to the Egyptian one.[881]

Transformations of the 3rd century AD

A lot of ink was spilled concerning the radical changes that took place in military command in the middle of the 3rd century, as the problem is difficult to solve. Quite naturally these changes are connected with the problem of the *primipilares*. Their role and place in banishing senators away from military command was established on two separate directions, somewhat resembling, but opposite from the point of view of the rank of the legate's replacement.

Praefectus legionis is a title bestowed to a knight commanding the legion, in the 1st century only in Egypt but from the 3rd century onwards for all legions from the Empire. Still, the same title is used in the 2nd – 3rd centuries to refer to a totally different officer. *Praefectus castrorum*, who initially was not an officer in the legion but supervised the camp, regardless of the number of units that stationed inside it, becomes a regular officer and his title is abbreviated from *praefectus castrorum legionis* to *praefectus legionis*. It seems thus that *praefectus castrorum* was put in charge of the Egyptian legions at the beginning of the Principate and of all others from Gallienus onwards.

Following Domitian and the separation of each legion in its own camp *praefectus castrorum* loses his independence, becoming a rank in the legion, also determining the gradual change of the title to *praefectus legionis*. The disappearance of the title *praefectus castrorum* is also to be blamed on Severus' official decision. Judging from inscriptions containing both titles we notice the change took place sometime during the 2nd century but we cannot pinpoint the exact date. Severus did not involve so deep in organizing the ranks that he would modify a relatively low title. Also *praefectus castrorum* does not receive an *officium* only now, nor is it increased. To this sense other possible clues regarding the change in rank at the beginning of the 3rd century are circumstantial, the only change being the shortening of the title.[882]

It is generally accepted what Aurelius Victor says about senators being removed from military command by Gallienus, especially due to the fact that after his reign we no longer find any *legati legionis* mentioned. Also the inscriptions mentioning *praefectus legionis a(gens) v(ice) l(egati)* start to appear now.[883] The expression *agens vice legati* is obviously meant to emphasize the difference between the new commanders and the *praefecti castrorum*, who are now *praefecti legionis*. Another method of differentiation is the expression *v(ir) e(gregius)*, obviously not used for the *praefectus castrorum*; it is mostly encountered in Africa with reference to the commanders of the *III Augusta*. Although this reference does not become general we must assume that is was the explanation of the difference between the two ranks with the same name. And by excluding the two senators from the legion, the legate and the tribune, there was still one knight on a higher rank than the *praefectus castrorum*, namely the *primipilus II*.[884]

Replacing senators with knights at the command of the legions is not a punishment for senators, as Aurelius Victor believes because of the period in which he wrote. The command system created by Augustus had a great weakness: incompetence. The recognition of the senatorial *dignitas* by Augustus was one of the conditions for accepting monarchy, but the weakness of this system passed unnoticed for almost 300 years. Knights on the other hand are better suited for these posts. They do not have claims about Republican traditions, they accept the idea of monarchy, are more loyal. Their career depends more on their performance and fidelity than on a *cursus honorum*. From Hadrian onwards we have two types of equestrian careers: a military and a civilian one, leading to some exceptional military specializations. Severus continued down this path, promoting even soldiers into the equestrian order. Thus a multitude of very experienced knights occurred, prepared to command legions. This seems to have been Gallienus' logic, who assumed that after repeated failures to maintain order and the allegiance of the legions on the part of senators, experienced knight commanders might improve the situation. Equestrian officers were quite competent. The lack of experience of the young ones is compensated by those who become knights at an older age, after a long service and command in the legion.[885]

Many such officers, former *praefecti castrorum* join the *ordo equester* especially in the 3rd century. The origin and evolution of the post of *praefectus castrorum* has been

[880] Devijver 1989, 29.
[881] Domaszewski 1908, 120-122.
[882] Keyes 1915, 21-25.
[883] CIL III 3424 - a. 267.
[884] Keyes 1915, 34-38. The same opinion is supported by Dobson, Domaszewski-Dobson 1967, xxxii; Dobson 1974, 429-431 and Dobson 1978.
[885] Osier 1977, 678-680.

discussed for a long time. It is certain that, towards the end of the 2nd century, the name gradually loses ground to *praefectus legionis*, who still places third in the legion's command hierarchy. In spite of the custom that these prefects could take command of the legion while senators are absent, Gallienus no longer appointed senators, automatically placing these prefects in command.[886] We notice how in this case the source is interpreted in two ways in order to obtain different conclusions.[887]

Under Aurelian the term *praefectus legionis* was already perpetuated and generalized meaning there also was a *cursus* including this post. Vegetius[888] seems to support the coexistence of *praefectus castrorum* with *praefectus legionis*, but he is not a reliable source for these matters. The term *agens vice legati* does not differentiate between the two types of *praefecti legionis*, because it also appears in the Parthian legions, but serves to maintain a façade of tradition, to maintain some of the seniority of the senators.[889]

It has been suggested that the new prefects are promoted from the ranks of the *praefecti castrorum* but this cannot be generalized, certainly there are variations. The official title of the commander was *praefectus legionis agens vice legati*, in order to include the whole list of prerogatives of the former senator.[890]

The situation remains unclear for the next 100 years as we can well see. Every one of these theories has their own logic and arguments. First of all promoting the *praefectus castrorum* to the command of a legion he already knows and works with, perhaps after tens of years of service, only recognized his right at being the third officer in the legion. Herein lays the problem, because the third officer of the legion seems to be the *primipilus II*. After completing the equestrian *militiae* in Rome he had gathered enough military and administrative experience to be able to command a legion. This was in fact the practice in the legions from Egypt. The system was probably the same in the case of the three *Parthicae* legions. There is no reason why this model cannot be applied to most legions, then, starting with Gallienus, to all legions. Perhaps we should look for the role of the *praefectus castrorum* more in the administrative and functional registry of the legion and only in exceptional cases in the military command one. In fact this would be the ideal way to organize a legion. One of the *primipilares* would provide administrative command for the legion while the second, with his experienced improved by the three tribunates held in Rome, would contribute to command and tactical leadership.

To this knowledge we should add two recently published inscriptions from Potaissa, statue bases for Fortuna and Aesculapius and Hygia, discovered in the baths of the fortress. They are dedicated by a M. Publicianus Rhenus, who bears a title attested in premiere: *praef(ectus) alae I Bat(avorum) ∞ agens vice praef(ecti) leg(ionis)*.[891] Dated under Gallienus, by the first author, the monuments seem to be slightly earlier, perhaps under Aemilius Aemilianus (AD 253).[892] However, this is not the most important piece of information coming from the inscriptions. The title, revealing a double command of a knight, that of an *ala milliaria* and that of a legion, sheds a ray of light upon the hierarchical position of the *praefectus legionis*. It basically states that the prefecture of a legion could be held only by an equestrian *a militiis*. This is yet another proof that inclines the balance towards accepting the *primipili II* as 3rd century *praefecti legionis*, as they already served in the three/four urban *militiae*. On the other hand, it could be that simply the vicinity of the fort of Războieni-Cetate to the fortress of Potaisssa was the reason for this irregular appointment. But the assumption that bot the *ala* and the legion shared the fortress is however not likely.[893]

Equestrian careers of the *primipilares*

The tribunate of the cohorts in Rome is held only by *primipilares*. They were vital for the beginnings of their careers although the post of *tribunus cohortis urbanae* appears only starting with Claudius and that of *tribunus cohortis vigilum* starting with Nero. The fact that these posts are reserved for *primipilares* is not connected with the open path towards procuratorships but with the importance of commanding all military and paramilitary troops from Rome. When the *equites singulares Augusti* come to Rome they will also be commanded by a *primipilaris*, just like the legion *II Parthica*. These commanders needed to be completely loyal and the emperor needed to be able to count on them. Not all *primipilares* become tribunes in Rome, where there are only seven positions as tribune of the *vigiles*, four as urban tribune and ten as praetorian tribune, all of them to be occupied for one year. The tribunes of *equites singulares* and of the urban cohorts outside Rome held their posts for a longer period and they enjoyed other privileges. We cannot determine a fix promotion scheme between the tribunates in Rome because it depended on what posts were vacant. A post might be omitted if several superior posts were open.[894]

Perhaps 10 out of the 30 *primipilares* that graduated yearly from the legions were chose for tribunates in Rome. Considered separately it seems that the *tribunus cohortis vigilum* became vacant quite often. In the vast majority of known cases one becomes *tribunus vigilum* right after his primpilate. Out of the 24 tribunes for whom we know their previous careers, 17 became tribunes precisely after finishing their primpilate. The first generations of tribunes of the urban cohorts also obtain this post directly after the primpilate, not coming after the

886 Osier 1977, 680-681.
887 *Contra* Keyes 1915, 29.
888 *Epitoma* 2.9-11.
889 Osier 1977, 682-683.
890 Southern 2001, 92-93.
891 ICLP 22-23.
892 Information I. Piso.
893 For the entire theory see Bărbulescu 2012, 186-212.
894 Dobson 1978, 75.

tribunus vigilum and praetorian tribunes previously occupy other equestrian *militiae*. That is why we have no reason to assume the existence of such a commander before the time of Claudius (he is not mentioned in inscriptions anyway). After Claudius and until the middle of the 3rd century the general rule was to accede to the tribunate of a cohort of *vigiles* right after the primpilate. A change seems to have taken place only in the 3rd century. Studying three post-Severan careers, all containing the post of *tribunus vigilum*, we notice that only one of them is promoted directly from *primipilaris*, the other two having completed the tribunate *equites singulares Augusti* or *praefectus castrorum*.[895] The tribunate of the *vigiles* is in fact used to gain access to the other two.[896]

Evidence concerning service as *tribunus cohortis urbanae* is rare, more common for those outside of Rome. One of them apparently held the post for four years and all examples fail to mention the tribunate of the *vigiles*, as well as the praetorian one or *primipilus II*. Apparently the tribunate of the urban cohorts outside of Rome comes immediately after the primpilate.[897] The tribunes of the urban cohorts are the direct superiors of the centurions, and the officers commanding these troops. Their rank is second in the hierarchy of the tribunates from Rome. Starting with Claudius they regularly promote to praetorian tribunates. All of them are *primipilares* and later will serve in important procuratorships and prefectures. Analyzing the careers of a few important characters, such as Q. Marcius Turbo,[898] we can notice that these tribunates lasted three years, each of them being held for year. Still there are no clear rules, especially concerning the tribunates of the *vigiles* or *equites singulares* that can last up to three years.[899]

Tribunus equitum singularium seems to be a very important post, similar to the praetorian tribune, especially because the owner of the office had daily access to the emperor. It was usually held after the tribunate of the *vigiles*. Apparently it was held for 2-3 years, in any case longer than the other tribunates, but the delay was eventually picked up afterwards, as we can see from the careers of those who served on this post.[900] In the 2nd century, when the *equites singulares Augusti*, were camped in a single fortification they had a single tribune, after Severus doubled their numbers from 1000 to 2000 and placed them in two fortifications, two such tribunes were needed. They are assisted in their multiple duties by a staff or organized petty officers. Iconographic evidence shows a connection between them and the emperor on the battlefield. In the 3rd century they lose their status and rank and are recruited from lower social classes.[901]

Each praetorian cohort was commanded individually by a *tribunus* as early as the Julio-Claudian period. To get the qualification needed for the praetorian tribunate one had to have held a post of *primus pilus*, then both the tribunates of *vigiles* and *urbaniciani*.[902] *Tribunus cohortis praetoriae* was probably held for one year. After completing the praetorian tribunate, a *primipilaris* could return to the provincial army as *primipilus II* or follow an equestrian career.[903]

The recruitment method for tribunes is the same like the one of the *primi pili*. No details are offered about the career prior to the primpilate. Direct promotions are available only to those from the Praetorian Guard; legionaries are almost nonexistent. The preference for Italians during the Antoninus Pius – Commodus period is the easiest one to notice. It can be said that the choice between a career in Rome and one as *praefectus castrorum* also depended on the *origo*, thus there is a differentiation even between brothers.[904]

A number of 72 tribunes give no information concerning their careers after the tribunate. Besides the agitated periods from the beginning of the Principate and the 3rd century, they do not die as tribunes. This might indicate a majority who go towards a procuratorship, not as the beginning of a career but maybe as a reward. Out of the ten former tribunes and *primipili II*, one or later five await the command of a legion, in Egypt, or Mesopotamia or Italy, the rest having to compete with the knights for the 14 posts of *procurator ducenarius*, five of them being perhaps vacant every year. Thus some are tribunes or *primipili II* for a longer period of time, waiting for a vacant post. The tribunes from Rome are more than commanders of the cohorts. They must represent the emperor's authority in Italy, like Staius, who is sent to apprehend the leaders of a slave rebellion.[905] Others maintain public order in Rome, as executioners,[906] torturers[907] or assassins.[908] Some of them represent persons authorized in solving border disputes[909] or are sent as tax collectors.[910] Their ability to be dependable offered them an open door for promotion and

[895] Sablayrolles 1996, 141-143; Dobson 1974, 414 and note no. 95.
[896] Keppie 1984, 159.
[897] Dobson 1978, 76.
[898] Concerning the replacement of the tribunate of the *vigiles* with that of the *equites singulares Augusti*, in Turbo's career, see also Blommaert 1989, 306-307.
[899] Freis 1962, 81-84.
[900] Dobson 1978, 77-78.

[901] Speidel 1994, 95-102.
[902] Keppie 1996, 113.
[903] Keppie 1996, 113; Dobson 1978, 79-80. A virtual promotion scheme would look like this: seven *primipilares* become *tribunus cohortis vigilum*, after one year four of them move on to *tribunus cohortis urbanae*, one, the most talented, advanced to *tribunus equitum singularium Augusti*, the other two probably become *tribunus cohortis urbanae* outside Rome, if the posts were vacant. *Tribuni cohortis praetoriae* are chosen from the ranks of the *tribuni cohortis urbanae* and occasionally from the *tribunus equitum singularium Augusti*. It is obvious that it was not mandatory that each post be held for only one year.
[904] Tacitus, *Hist.* 2.26; *Ann.* 15.69. Dobson 1978, 82-83.
[905] Tacitus, *Ann.* 4.27.
[906] Tacitus, *Ann.* 15.61-67.
[907] Cassius Chaerea and Josephus, *Antich. Jud.* 19.34-36.
[908] Tacitus, *Ann.* 13.15.
[909] CIL X 1018.
[910] Suetonius, *Calig.* 40.

favors, like in the case of all *primipilares*. No other circle owes so much to the emperor, so they will serve him forever. Even though many of them do not have a complex career after the primpilate, they come in Rome to finish their career in a glorious manner, after 20-30 years of service. For others it was an important moment, when it was decided whether they were ready for greater responsibilities.[911]

After serving as tribune of a praetorian cohort followed the **iteration of the primpilate.** Inscriptions do not mention the legion of origin because the two primpilates are associated, before or after the mentioning of the praetorian tribunate. Even if the legion is mentioned in the case of the first primpilate it is no longer so for the second one. If the *primipilus* is tasked, since the time of the Republic, with the command of auxiliary troops the iteration takes place only in those cases when the men have proven their abilities by completing all equestrian *militiae*. The *primipilus bis* is included in the war council[912] or the general's staff since the time of Polybius but later on, during the Empire, he is removed from the command of the *centuria* so that he can become tactical assistant to the legion's legate. He thus held a position of a general's chief of staff; that is why only the officers with the most combat and command experience received this post. This structure of command would bring a unique character to the army. After completing the three tribunates the man was not downgraded, on the contrary.[913]

The position of *primipilus II* appears sometime under Caligula or Claudius. At the beginning of the Principate we still encounter the *primipilus bis* or *ter*, but Claudius changed everything with the creation of the *primipilus II*.[914] It is the access gate to procuratorships and follows automatically after the tribunate of a praetorian cohort even before the other two urban tribunates are created. It is very rarely omitted from the presentation of a career, only as an exception, and in many cases there are *primipili II* without any legion being mentioned. They are sent to Rome with different tasks then promoted, two of them remarkably as *praefecti legionis* in Egypt. Inside the legion, the *primipilus II* does not command the *centuria* because he already was *tribunus cohortis praetoriae*; he is situated between the two categories of tribunes, enjoying a greater authority than that of equestrian tribunes,[915] just like the *praefectus castrorum*. Still, it is not the same thing because the second one occupies his post right after the primpilate while the *primipilus II* already finished the equestrian *militiae* in Rome. It was not usual to task people of different ranks with the same duties, like in modern armies. The major difference was probably the salary, varying in between 100,000 and 200,000 *sestertii* from Domitian to Severus. This post cannot be easily explained away as an option to stay in the service while awaiting for a procurator's post to become vacant[916] and its attribution suggest duties of a more technical and administrative nature, in order to prepare the future procurators in this direction.[917]

The *primpilares* – besides commanding the Egyptian legions – competed with the members of the equestrian order for access to **procuratorships.** It is easier to notice patterns in the careers of *primipilares* promoted procurators than in the case of the other knights mostly due to their 30-40 years of military background. They began with a financial procuratorship of the 4th ducenarian rank. They could end their career here or advance to rank 3, to a presidial procuratorship. Rank 2, involving posts in Rome, was usually skipped with two exceptions: those who get here after commanding the legions in Egypt and must get administrative training and those favored by the emperor to be future *praefecti praetorio* or *vigilum* and who were kept close. They usually held the post of *procurator ludi magni*. Then some of them completed their career with an important procuratorship in Gaul, the prefect of the Ravenna fleet went to Misenum and thus finished his career, unless chosen for the more important prefectures. The important prefectures of the *vigiles* and of the *praetorium* are equaled only by the *praefectus annonae* in Egypt and the other secretarial posts are less suited for these *viri militares*.[918]

In this respect there seem to exist three types of procuratorian career accessible to the *primipilares*: a) direct promotion to a ducenarian procuratorship, followed by the end of the career; b) promotion to *praefectus legionis* in Egypt, than to a financial procuratorship followed by a presidial one; c) promotion to an important prefecture – men with exceptional careers. All of these careers start with a ducenarian procuratorship, the *primipilares* missing only a few special secretarial posts and this should not come as a surprise since those were obviously filled by civilians. Under Marcus Aurelius the number of presidial procuratorships drops, after the transformation of Noricum and Raetia in provinces with legions.[919]

We must mention that the *primipilares* represented the elite of the 2,000 centurions in the Empire. To be more precise less than 2% become *primi pili*, then many more of them must wait for vacant positions. That is why those who get far must be admired; certainly they represent trustworthy men for the Empire.[920]

[911] Dobson 1978, 86-87.

[912] However he is completely different from the *primipilus II* from the time of the Principate. If the primpilate is held twice in succession the term *primipilus bis* is used (during the Republic and at the beginning of the Principate), if it is intercalated by other commands then they use *primipilus II*. Because the first primpilate offers access to the command of a cohort from Rome, the tribunate in a praetorian cohort is followed by *primipilus II*. Allen 1908, 7.

[913] Domaszewski 1908, 114-115.

[914] Webster 1969, 118; Richier 2004, 498.

[915] Le Bohec 1994, 44.

[916] Dobson 1978, 88-91.

[917] Richier 2004, 498.

[918] Dobson 1979, 199-200.

[919] Dobson 1978, 105-108. He divides careers in three categories: A. presidial procurators, B. prefects of the fleet, C. superior prefects.

[920] Dobson 1978, 113-114.

III.1.3. *Primipilares* in religion, economy and society

The governor, the legion's legate, the tribunes, only as a group, *primus pilus* and centurions, only as a group, have the right to set up altars in the official places from the legionary fortress. The curator of the eagle – *aquila sancta*, is the *primus pilus* who can access the *aedes* and consult the auspicials. The connection between the *primus pilus* and the eagle goes beyond religion and there are many cases when *primi pili* risk their lives in order for the legion to escape the shame of losing their symbol.[921] We find evidence for this role of the *primus pilus* only until the reign of Severus, when the *haruspex* becomes an *officialis* of the governor and takes over this role.[922]

The *primus pilus* also had a religious role during military service, given that he was chief-centurion of the legion and was given the eagle and the *signa*,[923] meaning the organizing of the legion's religious life. As a privileged officer he could dedicate official monuments for the health of the emperor in the name of the legion, just like the governor or the legion's legate could. Altars or statue bases are placed in public places, in the courtyard of the *principia*, the oldest being the closest to the *aedes*. Official dedications of the *primi pili* are connected with the *aquila* and the emperor, are placed in the *principia* and the legate of the legions ensures they are set up. They are regularly built at one-year intervals, usually by the *primus pilus* who ended his office. This is a sacred date and probably corresponds to the date when the legion was created, *natalis aquilae*. Other inscriptions mark imperial victories and differ from one legion to the next.[924] Their number greatly surpasses that of inscriptions set up by other officers.[925]

The deities invoked in these monuments are not regular, with three exceptions – Mogontiacum (*honos aquilae*),[926] Novae (*bonus eventus*)[927] and Lambaesis[928] (Emperor's cult).[929] The setting up of these monuments is due to a competition generated on one hand by the legion's religious tradition and on the other by the image of social success and the desire to leave the most imposing monument for the legion he just 'led'.[930]

In order to establish the exact social role and status of the *primipilares* we must discuss first of all his payment, *praemia* and *dona*. The centurion was encouraged to aspire to the primpilate and thus to serve the emperor for as long as possible. One would become *primus pilus* at around 50 years of age although there are exceptions on both sides. Given that a centurion's length of service is about 20 years and his total military record is about 40 years, apparently 90 legionary centurions could be promoted each year. *Primus pilus* held that post for one year, meaning there are 90 candidates for approximately 30 posts. According to statistics, the chances for a centurion who survives 20 years of service to obtain the primpilate are about 50%. Still the emperor was not worried. After finishing his one year service a *primus pilus* had three options: retire with a *praemia militiae* of 600,000 *sestertii*,[931] remain in service as *primipilaris* in the *numerus primipilariorum*, in Rome or in the provinces, with the possibility of becoming *praefectus castrorum legionis* and finally, remain *primipilaris* in order to hold the tribunates of the troops stationed in Rome. In any case he joined a very select group. Out of approximately 600 *primi pili* and *primipilares* alive at a given time, 30 were *primi pili* in the legions, 30 were *praefecti castrorum*, 24 were tribunes in Rome,[932] one was *praefectus castrorum* in Egypt and 5 to 10 were procurators.[933]

In 13 BC, a legionary who withdrew from the army received 20 times his yearly pay (about 3,000 *denarii*), and a praetorian 10 times (about 5,000 *denarii*). The salary of the centurion was calculated in connection with that of the praetorians meaning the *praemia militia* is the same, 10 times the salary. At the time of Augustus's death this meant 37,500 *denarii*, respectively 75,000 for the *primi ordines*. Because of these impressive numbers length of service for centurions was extended. Caligula reduced this rate to just double the salary and the salary of the *primi pili* to 60,000 *sestertii*. An age limit was thus established for occupying the primpilate; it was 49 years of age under Vespasian. This was connected with the old manner of serving in the army, when service ended at the age of 45.[934]

Based on a passage from Juvenal,[935] we can say that the primpilate had its own rewards but only after a long service.[936] Even though during the time of the civil wars it seems profitable to have this rank, the connection with wealth must have occurred only after Augustus instituted the *praemia militia*. Caesar's centurions were also rich; he borrowed money from them so that he could wage a civil war,[937] but there was no central fund from where

[921] Tacitus, *Hist.* 3.22.4. Stoll 2007, 458.
[922] Domaszewski 1895, 111.
[923] Vegetius 2.8; Tacitus, *Hist.* 3.22.
[924] Dobson 1978, 155-160; Kolendo 1980, 49; Popescu 2004, 275.
[925] Kolendo 1980, 49.
[926] Herz 1975, 182-190.
[927] Kolendo 1980, 50-59.
[928] CIL VIII 2533, 2535, 2542. Herz 1975, 192-193.
[929] Popescu 2004, 275.
[930] We must include in the category of official documents the altar dedicated by L. Calvisius Secundus to Liber *pater*, for the health of Commodus, in AD 183-185, at Apulum. The official character of this act is underlined by the presence of the governor and the legate of the legion in the inscription. The worshipping of Liber *pater* in military environments is not unusual, similar situations being found at Aquincum, Oescus or Novae. Another monument, dedicated to Serapis, has a very personal character. The choice of divinity can be connected with the evolution of the dynastic idea at the end of the 2nd century but, in any case, he is the only *primus pilus* in the Empire who makes a dedication to Serapis. In their private endeavors *primi pili* set up dedications according to their own wishes. Popescu 2004, 276.
[931] Suetonius, *Calig.* 44.
[932] The salaries of tribunes from Rome: *tribunus cohortis vigilum* = *praefectus castrorum* – 80,000 *sestertii*/year, *tribunus cohortis urbanae* – 100,000 *sestertii*/year and *tribunus cohortis praetoriae* = *primipilus II* – 120,000 *sestertii*/year. Devijver 1974a, 76-77.
[933] Dobson 1970, 102-105.
[934] Domaszewski 1908, 118-119.
[935] 14.197: *ut locupletam aquilam tibi sexagesimus annus adferat.*
[936] Dobson 2000, 139.
[937] *BC* 1.39.

such a large sum of money could be paid. The title of *primipilaris* could not be inherited; it represented – just like the title of *consularis*, but on a different scale – a top rank obtained in one's career. It was the supreme glory for the legionary centurion and the money supported this position. This military aristocracy was more than just another rank beyond the legionary centurionate; the primpilate became the normal goal for every centurion serving in the legions or in Rome.[938]

Although the *praemia* of a centurion was large, that of a *primus pilus* was four times larger, the only downside being the advanced age.[939] We must assume though that the majority of centurions did not reach the primpilate.[940] The sum received by a *primus pilus* upon discharge was probably 600,000 *sestertii*, connected to a payment of 60,000 *sestertii*/year, perhaps increased to 100,000 by Domitian. According to M.A. Speidel, after all the salary indexations from the time of the Principate, the *primus pilus* will earn 432,000 *sestertii*/year under Maximinus, with a ratio of 1:60 to the legionary's pay.[941] It is certain that the *primus pilus* received double the payment of a *primus ordo* and four times as much as the other centurions so, under Augustus, he received upon discharge 600,000 *sestertii*, one and a half times a knight's census.[942]

It was still not possible at the beginning of the last century to decide in what social category to place the *primipilares*. In spite of clues given by literary sources, including those of the poets Ovid and Martial who claim *primipilares* are equivalent with knights and the reception of this social status once the primpilate was over, the general German-French belief was that the *primipilares* belong to the third Roman social class, the plebeians.[943]

With the same evidence however we can reach different conclusions. A number of inscriptions[944] were believed to indicate plebeian status for the *primi pili* because of the association with the expression *ex equite Romano.* However this proves the status of the person before enlistment and what matters is that he is still in service and has not received the *praemia*. *Primipilares* fulfill all conditions for access in the equestrian order: free birth, citizenship and money requirements, thus they can be considered knights. We do not know whether they are *eques publicus* or listed in the *decuriae iudicum*, but the son of a *primipilaris* is considered *eques publicus*. This would be very strange if the father is denied knight status. The fact that the fathers of these *equites* are referred to only as *primipilares* and not *equites* supports the custom of automatically considering a *primipilaris* to be a knight, no further proof being needed.[945]

The centurions are not subjected to monetary requirements in order to join the knightly order and evidence places them among the plebeians, just as the same evidence places the *primipilares* among the knights.[946] An example is the civilian career of the *primipilaris* Arrius Clemens who holds an office denied to the plebs - *curator reipublicae*. Thus it was superfluous for a *primipilaris* to mention his equestrian status in inscriptions; it is why it was often excluded. The system still functions under Augustus and separates two social classes in the army.[947]

We cannot say much concerning the social origin of the *primipilares*, except in the case when they are drawn from equestrians. We can see that they come from respectable families, many of them with a military tradition dating back a few generations, from Italy or provincial colonies. They do not have the opportunity for *adlectio* into the senatorial order, with only a few emergency exceptions, because they no longer have time (they are old) and they have their own specific career. Their descendants become senators at the second or even the first generation. The sons that did not become senators had two options. They could join the army, directly appointed centurions and also become *primi pili* or they could start an equestrian career and be happy with their status as knights. When a knight refers to his father being a *primipilaris* it does not mean that he did not become a knight, he simply adds a distinction. There are also the so-called military families, with brothers of the *primipilares* serving as centurions or also becoming *primipilares*.[948]

Also the knights who join the army can have brothers that follow an equestrian career, proving that centurions cannot be considered a group of plebeians. The relations between these orders exist even higher. Marriage takes place inside the caste of centurions - *primipilares*, though there is one case when the wife was the sister of a consul.[949] Finally we must observe that the second generation of the family of a *primipilaris* is completely absorbed in the equestrian order, thus any sign of its origin is lost, except for some inscriptions from the East that refer to old ancestors. The *primpilaris* has no other distinctive titles. Tacitus refers to him as *vir militaris*, which he is in fact. Later on the term *vir egregius* is used; this term seems to differentiate between knights with military origins and the simple *eques Romanus*. In the local administration the *primipilaris* has a tendency to occupy only higher magistracies, sometimes in the cities of origin. Other times they are *praefectus pro IIviris,* from the part of a member of the imperial family until the 3rd century when they are asked to become patrons and

[938] Dobson 2000, 140.
[939] Juvenal 14.197; Plinius, *Nat. Hist.* 14.9.
[940] Junkelmann 1986, 113.
[941] Speidel MA 1992, 101-104, especially 102. A ratio of 72:1 between the payment of the *primus pilus* and that of the legionary is also considered real but the sums remain the same, Herz 2007, 308.
[942] Dobson 2000, 142, Patterson 1993, 97. For decorations see Domaszewski 1908, 117-118; Maxfield 1981, 202-207.
[943] Schmidt 1886, 590.
[944] For example CIL III 750.
[945] We can even claim that the primpilate was a distinction for a knight. *Contra* Schmidt 1886, 591-593.
[946] Keppie 1984, 151.
[947] Schmidt 1886, 594-596; Alföldy 1988, 124.
[948] Dobson 1970, 107-109.
[949] Minicius Iustus – Plinius, *Ep.* 7.11.4 ;Tacitus, *Hist.* 3.7.

thus dominate municipal magistracies. Next to the veteran centurion, a *primpilaris* had the money and skills to hold municipal magistracies but both were excused from occupying them.[950] Moving up from the municipal level he could even lead the *concilium provinciae*, especially in the East.[951]

The newly formed cities from Italy benefit from a high flux of high-ranking veteran colonists (praetorians, *primipilares*) but also from the emperor's generosity, who assists them in setting up grandiose monuments. Centurions and *primipilares* will best exploit these new resources in order to reach the local or even central aristocracy.[952] Nero, who was born in Augustus's villa from Antium, founded a *colonia* there, where he settled many high-ranking soldiers, former praetorians or *primipilares*, according to Tacitus and Suetonius.[953] Selecting important officers to serve as leaders of newly created citizen communities continued the Republican custom of settling veterans *cum tribuni et centurionibus*. Nero had, in Rome, at his disposal, a number of such *primipilares*; their wealth was proverbial and this turned them into a valuable addition for every city. It is also believed that the whole theatre from Antium was built by these very rich *primipilares*.[954]

Their loyalty and debt to the emperor is proven by three different excerpts, where the emperor imposes monetary obligations on the *primipilares*, payments for spectacles or even forced colonization in an area that had to be romanized.[955] The *primpilaris* was a dominant figure in local and municipal administration and society had high expectations from him, importance and social prestige being obvious.[956]

Although it is a characteristic of the 3rd century, and later on of the Dominate, the economic role of the *primus pilus* and *primipilaris* is evidenced in several cases. Two altars discovered at Carnuntum[957] mention a *conductor* of a *pratra* in the *lustrum* of a *primipilus*. It was believed to be evidence of the origins of peasant-soldiers from the 3rd century. In fact it is another system, *lustrum* being more recently considered as a period of lease of the legion's territory and its means of productions. The *lustrum* seems to be the system where *conductors* appear and integrate themselves. There are examples for *lustrum primipili* or of superior officers. All inscriptions mentioning *lustrum* are dated between AD 205-244 so we could consider this to be an innovation of the Severan dynasty. It had some kind of personnel, including *signifer*, *mensores* and *conductores*, *dispensator*, the freedmen of the *primus pilus*, who sometimes call themselves *agentes in lustro*. Besides, them, the *actors* were mostly slaves or freedmen of the *primus pilus* not employees of the legions; they are private men who answer to their master.[958]

The *primus pilus* holds a separate accountancy for the *lustrum*, as shown by Vegetius,[959] mentioning a *commoda primipili*. This is not necessary an illegal income but an additional one, the profit of the *primus pilus*. *Lustrum* was not a private affair although the private personnel of the *primus pilus* could manage its administration. *Administratio primipili* meant the whole *lustrum* business, including profits (*commoda*) and losses/debts (*reliqua*). The *commoda* is mentioned only until the time of Gallienus, after this only the debts and the duties of the primipilate remain.[960]

We have other examples from other parts of the Empire. A novel altar from León mentions a certain Minucus Priscus, *primipilus legionis VII Geminae*, and one *Verna actor* who set up the monument. The problem is raised by this *actor*'s status, which may or may not be a slave, either a private one or belonging to the legion. *Lustrum primipili* becomes a special service of the *primipilus* 4 or 5 years long and could indicate a change in the length of service for the primpilate in the 3rd century. We cannot say whether this is a public or a private slave but we can assume that under Severus the *primus pilus* could hire private slaves to help him with his new accounting duties.[961]

The modifications from the 3rd century transformed the nature and functions of the *primus pilus*, by taking away his tactical duties.[962] Thus, gradually, he is more and more associated with the administration and the finances and there is even a tax named after him, *primipilum*. The legion still had a *primus pilus* until Diocletian but he had taken on many administrative tasks since the time of Severus-Caracalla. Around AD 286 the status of the *primipilaris* was mandatory and hereditary and will eventually disappear once its signification was lost in the new *field army*.[963] In any case, the position is gradually demilitarized and thus stripped of importance.[964]

950 *Digest* 49.18.2.5.
951 Dobson 1970, 110-114.
952 Patterson 1993, 109.
953 *Nero* 9: *Antium coloniam deduxit ascriptis veteranis ex praetorio additisque per domicili translationem ditissimis primipilarium.* Wesch-Klein 1998, 199-200.
954 Keppie 1984a, 86-88.
955 Suetonius *Tib.* 37, *Calig.* 38 and *Nero* 9.
956 Dobson 2000, 151-152.
957 CIL III 14350 and ILS 9103.
958 Mócsy 1996, 312-314.
959 *Epitoma* 2.7.
960 Mócsy 1996, 314-315.
961 Le Roux 1992, 267-268.
962 Richier 2004, 497.
963 Leaves room for the titles of *primipilus protector* or *primipilaris protector*. Strobel 2007, 271, 273.
964 Dobson 1974, 431.

III.2. *Primi pili* and *primipilares* from Dacia

A total of 25 *primi pili* and *primipilares* are active in different fields in Dacia. Obviously the *primi pili* are connected to the history and activity of the legion while the *primipilares* are also connected to the command of the garrison, thanks to the post of *praefectus castrorum* and civil life – for those of them who chose such careers. A number of 17 *primipili* and *primipilares* are attested epigraphically as part of the ***XIII Gemina*** legion while it stationed in Dacia. Seven of them are attested as *primi pili* of the legion, while serving in the army, and the remaining eight are simply *primipilares* or hold posts that follow the primpilate, *praefectus castrorum* or *legionis*. Only five persons who connected their primpilate with the ***V Macedonica*** while in Dacia are known: three active *primi pili* – though in one case the dating of his career is uncertain – and three *primipilares*, none of them with exceptional careers. Finally, another three *primipilares* do not serve in Dacia, but they are somewhat involved in its history.

In regards of their **origin,** little can be said, as the pattern seems to be quite similar to that of the centurions'. However, statistically, the things are looking slightly better, because of the small total number of attestations. Of the 25 *primi pili* and *primipilares*, seven (25%) have their exact origin known: three come from Italy,[965] one each from Asia Minor,[966] Spain,[967] Dalmatia,[968] and for one we can deduce his certain origin in Apulum.[969] Only two of them are knights.[970]

Concerning the **previous career**, it is only for eight of them that we hold such information. Four of them held legionary centurionates[971] and the other four come from posts in Rome. Two of them come from the praetorian petty officers,[972] two of them are *trecenarii*[973] and the last one comes from a centurionate career in *castra peregrionorum*.[974] One of the legionary centurions was previously a praetorian with a couple of extraordinary *evocatus* ranks.[975]

For the matter of the ***primipilaris*** **career**, a series of evidence occurs. The first branch of the career, that of legionary prefect/prefect of the camp was chosen by no less than 11 *primipilares*.[976] From amogst them, two are uncertain, as datable under Gallienus. Other higher military commands include that of *praepositus classis praetoriae*,[977] *dux legionum*,[978] encountered in two cases and *tribunus praetorianorum*.[979] Two of the *primipilares* passing through Dacia reached also procuratorships or high prefectures, four such posts being attested.[980] Another three held only municipal careers, being honored under several forms, *patronus*, *pontifex* or *decurio*, while still in service.[981] Finally, three of them are satisfied with a quiet relaxing life and they are simply mentioned as *primipilaris*.[982]

[965] Nos. 3, 15 and 18.
[966] No. 13.
[967] No. 16.
[968] No. 23.
[969] No. 24. And possibly no. 7.
[970] Nos. 15 and 24.
[971] Nos. 6, 16, 18 and 23.
[972] Nos. 18 (afterwards becoming also legionary centurion) and 25.
[973] Nos. 14 and 15.
[974] No. 24.
[975] No. 18.
[976] Nos. 5, 6, 9, 12, 14, 18, 20, 21, 23 and 24.
[977] No. 23.
[978] Nos. 23 and 25.
[979] No. 25.
[980] Nos. 23 and 25.
[981] Nos. 16, 18 and 24.
[982] Nos. 4, 7 and 14.

	Name	Origin	Previous post	*PP*	Source	Place
3	L. Calvisius Secundus	Falerio		*XIII Gemina*	CIL III 1092=IDR III/5 235	Apulum
6	M. Ulpius Caius		*> III Italicae*	*XIII Gemina?*	IDR III/5 448, 436	Apulum
7	Aur. Sabinus	Apulum?		*XIII Gemina*	CIL III 1191=IDR III/5 506	Apulum
13	C. Iulius Candidus	Ancyra		*XIII Gemina*	CIL III 267=IDRE II 398	Ancyra
14	Sex. Vibius Gallus		*CCC*	*XIII Gemina*	CIL III 13648, 14187=IDRE II 389-392	Amastris
15	L. Aninius Firminus	Tridentum	*eq publicus* *CCC*	*XIII Gemina*	ActaMN 43-44/I, 179	Apulum
16	C. Iulius Lepidus	Ieso	*> XIII Geminae ?* *> I Adiutricis* *> X Geminae* *> XX Valeriae V* *> VII Geminae*	*XIII Gem?*	CIL II 4463=IDRE I 174	Ieso
18	M. Apicius Tiro	Ravenna	*> XV Apollinaris* *evoc a commentariis custodiarum* *optio* *evoc salariarius*	*XXII Primigenia*	CIL XI 19=IDRE I 135	Ravenna
23	L. Artorius Castus	Epetium	*> III Gallicae* *> VI Ferratae* *> II Adiutricis* *> V Macedonicae*	*V Macedonica*	CIL III 1919, 14224 IDRE II 303, 304	Epetium Salonae
24	P. Aelius Marcellus	Apulum	*eq Romanus* *subprinceps peregrinorum* *> frumentarius*	*VII Gemina*	CIL III 7795, 1181, 1182=IDR III/5 442, 441, 439 CIL XI 5215, 5216	Apulum Fulginiae
25	*Unknown*		*cornicularius praef praet*	*?*	CIL VI 1645=IDRE I 19	Roma

Table. III.1. Origins and previous careers of the *primi pili* from Dacia.

No. **3**, originating from Falerio, Italy, sets up a statue base in his official role as *primipilus* of the legion. We are dealing with an official dedication, from the part of the entire legion similar to the series discovered at Novae or Mogontiacum and previously mentioned. It was probably set up at the end of Secundus's primpilate as was the custom in the other cases as well. The monument is dedicated for the health of the emperor, Commodus, while the consular L. Vespronius Candidus was governor, under the care of the legion's legate, C. Caerellius Sabinus, so sometime between AD 183-185.[983] Only an official dedication can be so thoroughly edited and dated.

No. **13**, a *primipilaris* from the legion stationed at Apulum, could have returned to his native lands after completing his career or after he had taken part in Trajan's Parthian campaigns.[984] In any case he dies and is buried here, at Ancyra, by his heirs, according to his will.

No. **14** is the well-known Sex. Vibius Gallus, *praefectus castrorum* of the legion *XIII Gemina*, probably *primipilus* in the same legion, a former *trecenarius* – meaning he was three times centurion in the cohorts from Rome – who was decorated by two emperors, following a series of wars. The most recent research indicates them to be Domitian's and Trajan's Dacian wars,[985] dating a few of the monuments in AD 115. He has a typical career for obtaining access to the primipilate, following the centurionates in Rome – three in his case – then, his career seems to end with his post of *praefectus castrorum* of the legion that will remain in Dacia.

No. **15** is a *primipilus* of the legion, recently discovered and published, coming from Tridentum, Roman knight, who bears the title *trecenarius*. From his position as *primipilus*, probably during or at the end of his one-year service, he builds a *tetrastylus* and a silver *aquila*. This *aquila* is not the official standard of the legion but a silver replica, probably placed at the entrance in the *aedes* – the entrance is rebuilt with this occassion by this *primipilus* as a *tetrastylus*. The works are dated in the Severan period.[986]

No. **16** is one of those citizens that are too valuable to the community not to be enlisted as decurion, even while in military service. He originates in Ieso, Taracconensis, where we can find his career on his funerary monument. He had a rather long centurion career, passing through no less than five legions. Their order must be inverse, as the man is more likely to be enlisted in *VII Gemina*, practically at home, than in *XIII Gemina*, on the Danube frontier. If this assessment is correct, than he could have been a *primus pilus* of the Dacian legion, where he must have held his last centurionate. In any case, the first half of his career is placed in the West, afterwards passing on the Danube frontier. The *ordo decurionum* of Barcino chooses to honor him by enlisting him as decurion of the city, even while in service.[987]

No. **18**, originating in Ravenna, Italy, has also an interesting career. He probably comes from the praetorian *principales*, from where he is recalled under at least two exceptional posts: *curator ab indicibus evocatus salariarius* and *evocatus a commentariis custodiarum*. In between these two posts he is an *optio*, but we don't know of what kind, could be also of the *evocati*.[988] Finally he reaches the centurionate in *XV Apollinaris*, then the primipilate in *XXII Primigenia* and the career ends with the legion prefecture in *XIII Gemina*. The centurion and *primipilaris* career seem to have developed during or immediately after the Dacian wars of Trajan, when the three legions are together. This is also enhanced by the rather short centurionate career (only one post, probably between the *primi ordines*) before the primipilate. Back at home he is the patron of the Ravenna *municipium* and *pontifex*, again exceptional honors granted to military men still in service.

We already know no. **23,** L. Artorius Castus, from the group of centurions who also served in Dacia; he becomes *primus pilus* of the *V Macedonica* legion and he has a special career, because after the primpilate he holds a special command. He holds at least four legionary centurionates before becoming *primus pilus*, he then leads a vexillation from the praetorian fleet from Misenum as *praepositus*, after which he becomes prefect of the legion *VI Victrix*, a position from where he seems to have held an extraordinary command as *dux* of a vexillation of *Britanniciani* legionaries and auxiliaries. Finally he was appointed procurator of Liburnia, a region included in Dalmatia. Although we cannot be certain, his career might date back to the time of the Marcomannic Wars so we cannot establish with certainty if he served his primpilate in the army of Dacia.

No. **24** is P. Aelius P.f. Papiria Marcellus, a Roman knight with a distinguished military career. Thanks to his social position he joins the army directly as *centurio frumentarius*, in the *castra peregrinorum* from Rome. He is promoted in the same unit to *subprinceps peregrinorum*, second in command of the troops stationed here. From this position he passes to the legionary centurionate, ensuring for himself also the access to the primipilate. In the legion *VII Gemina pia felix*, he holds all three centurionate posts, *hastatus*, *princeps* and *primipilus*. Judging by the epithet of the legion we can date his career in the Severan era. It is again proof that promotion between legionary centurionate posts did not necessarily mean transfer between legions. After the primipilate he is *adlectus ad munera praeff(ectus/orum) legg(ionum) VII Claudiae et I Adiutricis*.[989] He probably didn't occupy both prefectures at the same time[990]

[983] Piso, IDR III/5 235.

[984] Moga 1985, 106.

[985] Dobson 1978, 223 and Maxfield 1981, 204.

[986] Moga, Piso, Drâmbărean 2008. 179-184.

[987] The centurionate in *XIII Gemina* is however not doubtless. It may have been just as well in *XIIII Gemina*, in which case his itinerary down the Danube would have stopped in Carnuntum. Petolescu, IDRE I 174.

[988] For these posts see Dobson 1978, 324, no. 245.

[989] Dobson 1978, 319.

[990] Dobson 1978, 319.

however the case seems to be similar to no. **6**, although not identical. His military career ends here. We can notice that in spite of his obvious membership to the equestrian order he does not enjoy a special career after the primpilate, being content with a classic post of *praefectus castrorum*, held twice, and afterwards probably concentrating on a civilian career. Also we can say that the only advantage his social status offered him is that he joins the army directly as a centurion in Rome, in this period in the *castra peregrinorum*, which obtains a great importance as a recruitment pool for professional military personnel starting with Severus. He certainly originated from Apulum because he is belongs to the Papiria tribe.[991] He has a prolific civil career, crossing across several provinces. He is decurion and patron of *colonia* Apulum and patron of three Italic cities: Fulginiae, Forum Flaminium and Iguvium. He holds the offices of *flamen lucularis*[992] and *sacerdos Laurentium Lavinatum*. From this position he donates food and money to the citizens of Forum Flaminium. He is a very important character, part of the provincial elite in both Dacia and Italy and in Apulum he is mentioned on at least three monuments. Here he adopts the son of another knight, *a militiis*, perhaps to gain favor for his career. It is a typical career of a *primipilaris* who is satisfied with provincial honors and is not interested in the high-ranking equestrian procuratorsips and who uses the primpilate probably for the payment, social position and prestige that it brings.

No. **25** from Rome does not mention the legions in which he served as centurion and *primus pilus* but these posts are followed by the command of a vexillation from the Dacian legions, as *dux leggionum Daciae.* He can be considered a *primipilus* of a Dacian legion because we have no contradictory information. Following the primpilate he completed the praetorian tribunate under Philippus Arabs, being afterwards rewarded with the posts of *procurator Lusitaniae*, *procurator ludi magni* and *praefectus vehiculorum*.[993] If he completed his primpilate in Dacia he is the only *primipilaris* from this province who follows such an equestrian career.

If no. **6** is the same person as a centurion of *III Italica* attested in Apulum,[994] then we have yet another *primipilaris* that offers details on his previous career. M. Ulpius Caius, *(centurio) leg(ionis) III Ital(icae) Antoninianae* dedicates a statue base to the governor of Dacia, L. Marius Perpetuus in Apulum. It seems that a few years later, he himself will have a statue dedicated to him, in which he is presented as *praefectus legionum IIII Flaviae et XIII Gem(inae)*. This formula, similar to that of no. 24 does not automatically imply that the two prefectures were simultaneous, although the idea is somewhat tempting. He probably arrived in Dacia for the completion of the primipilate, after which also the legionary prefectures occurred.

No. **5**, the legion's *praefectus castrorum* benefits from an honorific monument set up with public funding by the decurions and citizens living in the *canabae*. Although in the *canabae* military administration does not apply, it was preferable to have warm relations with the officers, especially one with significant administrative duties concerning the whole military installation from Apulum.

On the other hand no. **9**, *praefectus legionis XIII Geminae Gallienianae*, appears to be a prefect commanding the legion who maintained his command after Gallienus's reform. We cannot clearly classify his ranks because there is no mention *a(gens) v(ice) l(egati)*, or *v(ir) e(gregius)*, and his name is not one usual for an equestrian. In any case, since he was commanding the legion, he must have at least served his second primpilate and thus as tribune of the cohorts in Rome. He sets up a votive altar for the gods Aesculapius and Hygia, following the success of his treatment.[995]

No. **10** is mentioned on a collective altar dedicated to Mithras by several *agentes in lustro* of several *primipili*, at Aquileia. The *signifer* of the legion *XIII Gemina*, Valerius Valens, appears as *agens* in the *lustrum* of Aurelius Zeno who is clearly considered by B. Dobson to be the *primipilus* of the same legion.[996] No. **11** is also mentioned at Aquileia, as *primipilus*, on an altar dedicated to Mercurius *Augustus,* by the above-mentioned *signifer* and by an *actor*, Domitius Zosimos. The freedman, a former private slave of the *primipilus*, probably belonged to Terentianus's *lustrum* (called here *ratio*) who, judging by the precept from the other altar, can be considered *primipilus* of the *XIII Gemina*. It is dated, based on the names of the consuls, in AD 244.
No. **12**, *praefectus legionis XIII Geminae*, is considered at least by some,[997] to be a commander of the legion following Gallienus's reform but we have no real evidence for such suppositions.

No. **17** is one of the most recently discovered *primus pilus legionis XIII Geminae*. From his fragmentary monument, only his name and rank can be deducted. However, other clues of his existence in Apulum appear under the name C. Pomponius Fuscus,[998] a *IIvir* of the *colonia*, who dedicates two altars to Iupiter *optimus maximus* and to Aesculapius and Hygia. The analogy is a little far-reached, but the *nomen* and *cognomen* coincidence could point out to a common origin, if not also to ascendance.

No. **20**, prefect of the legion *V Macedonica*, under the common rule of Gallienus and Valerianus (AD 256-258), seems to be the commander of this legion given the military reforms enacted by Gallienus.[999] Still, if we date

[991] Dobson 1978, 319; Piso, IDR III/5, 442.
[992] Somewhat strange since this position is reserved for the plebeians.
[993] Dobson 1978, 304-305 and Petolescu, IDRE I 19.
[994] CIL III 1178=IDR III/5 436.

[995] Piso 2013, 354-355 considers him also a commanding prefect, excluding from this time period the existence of the *praefectus castrorum*, or his successor *praefectus legionis*.
[996] Dobson 1978, 342.
[997] Moga 1985, 29, 89.
[998] IDR III/5 16 and 160.
[999] Bărbulescu 1987, 65. Piso 2013, 353-354 dates more precisely the inscription AD 255-256, because of the legionary title.

Gallienus's reform in AD 260 or 262,[1000] then he cannot fit in this group, being a simple *praefectus castrorum*.[1001] From this position he completes the temple of the god Azizus *bonus puer conservator* in Potaissa, for the health of the imperial family and the prosperity of his legion/the legion under his command. Like other examples from this period it depends on the date of Gallienus's reform. However, at about this time it is difficult to believe that Dacia had a senator as governor, or legionary legates.[1002]

No. **21,** another *praefectus legionis V Macedonicae c(onstantis?),* sets up a votive altar for Iupiter *optimus maximus*, the other gods and the *genius* of the sanctuary. He is believed by most academics to be a *praefectus castrorum* from the time of Commodus.[1003]

No. **1**, a regular *primus pilus* of the legion *XIII Gemina*, dedicates the funerary monument for his wife at Apulum.

No. **2**, another such *primus pilus*, dedicates a votive monument to the god Serapis, one of the few such dedications in the official military milieu of Dacia. No. **4** is called simply a *primipilaris*, probably dedicating a votive monument. Since he apparently remained here after the completion of the primipilate, this could have happened in *XIII Gemina*. Another 'simple' *primipilaris*, no. **7**, raises the funerary monument of his wife. Due to his name (Aurelius, without *praenomen*) and burial of his wife (bearing the same name) in Apulum, I think it is safe to presume his origin in Dacia. The *primus pilus* of no. **8** is mentioned on the funerary monument that his daughter and a freedwoman erect for his wife/her mother/patron. This man either has made his family in Apulum, or he had brought it with him all throughout his career. Finally, both nos. **19** and **22** are *primi pili* of *V Macedonica*, in Potaissa, the last one being recently published as revealed in the centurion lists from the double dedication for Caracalla and Julia Domna. This *Tib(erius/urtius?*[1004]*)* is the first centurion in the lists, marked clearly as *P P* and his name written in larger letters than of the others centurions.[1005]

[1000] Alföldy 1988, 166.
[1001] Piso 2000, 217.
[1002] Piso 2013, 354.
[1003] Dobson 1978, 356; Piso 2013, 357.
[1004] A Tiburtius is centurion of *XXX Ulpia Victrix*. Reuter 2012, 94-95.
[1005] See Bărbulescu 2013, ICLP 10-11.

C	Name	PP	Military career	Unit	Eq./Civilian career	Source	Place
1	Sempronius Priscus	*XIII Gem*				AE 2004, 1203=ILD 460	Apulum
2	C. Cosconius Sabinianus	*XIII Gem*				CIL III 7769=IDR III/5 315	Apulum
3	L. Calvisius Secundus	*XIII Gem*				CIL III 1092=IDR III/5 235	Apulum
4	Aelius? Longinus	*XIII Gem?*			*primipilaris*	CIL III 7784=IDR III/5 376	Apulum
5	M. Ulpius Apollinaris	*XIII Gem?*	*praef castr*	*XIII Gemina*		AE 1910, 84=IDR III/5 438	Apulum
6	M. Ulpius Caius	*XIII Gem?*	*praef legg*	*IIII Flavia*		CIL III 1201=IDR III/5 448	Apulum
				XIII Gemina			
7	Aurelius Sabinus	*XIII Gem?*			*primipilaris*	CIL III 1191=IDR III/5 506	Apulum
8	L. Valerius Montanus	*XIII Gem*				AE 1983, 817=IDR III/5 581	Apulum
9	M. Aurelius Veteranus	*XIII Gem?*	*praef leg (a v l?)*	*XIII Gemina*		CIL III 1560=IDR III/1 54	Băile Herculane
10	Aurelius Zeno	*XIII Gem?*				CIL V 808=IDRE I 143	Aquileia
11	Domitius Terentianus	*XIII Gem?*				CIL V 8237=IDRE I 144	Aquileia
12	Q. Claudius Capito	*XIII Gem?*	*praef leg*	*XIII Gemina*		AE 1938, 55=IDRE II 362	Philippi
13	C. Iulius Candidus	*XIII Gem*				CIL III 267=IDRE II 398	Ancyra
14	Sex. Vibius Gallus	*XIII Gem*	*praef castr*	*XIII Gemina*	*primipilaris*	CIL III 13648, 14187	Amastris
						IGRR III 1434=IDRE II 389-392	Paphlagonia
15	L. Aninius Firminus	*XIII Gem*				ActaMN 43-44/I, 179	Apulum
16	C. Iulius Lepidus	*XIII Gem?*			*dec col Barcino*	CIL II 4463=IDRE I 174	Ieso
17	M. Pomponius Fuscus	*XIII Gem*				Lupa 19370	Apulum
18	M. Apicius Tiro	*XXII Primig*	*praef leg*	*XIII Gemina*	*patr mun Ravenna*	CIL XI 19=IDRE I 135	Ravenna
					pontifex	CIL XI 19=IDRE I 135	Ravenna
19	M. Statius Priscianus	*V Mac*				CIL III 910	Potaissa
20	Donatus	*V Mac?*	*praef leg (a v l?)*	*V Macedonica*		CIL III 875	Potaissa
21	Optatus	*V Mac?*	*praef leg*	*V Macedonica*		CIL III 892	Potaissa
22	Tib(erius/urtius?)	*V Mac*				ICLP 100-11	Potaissa
23	L. Artorius Castus	*V Mac*	*praepositus*	*classis Misen*	*proc Liburniae*	CIL III 1919, 14224	Epetium
			praef leg	*VI Victricis*		IDRE II 303, 304	Salonae
			dux	*leg II Britann*			
24	P. Aelius Marcellus	*VII Gem*	*praeff legg*	*VII Claudiae*	*flamen lucularis*	CIL III 7795=IDR III/5 442	Apulum
				I Adiutricis	*sacerdos Laur Lav*	CIL III 1181=IDR III/5 439	Apulum
					dec col Apulensium	CIL III 1182=IDR III/5 441	Apulum
					patr col Apulensium	CIL XI 5215, 5216	Fulginiae
					patr civ Foroflamin		
					patr civ Fulginiatium		
					patr civ Iguvinorum		
25	*Ignotus*	*?*	*dux*	*legg Daciae*	*proc Lusitaniae*	CIL VI 1645=IDRE I 19	Roma
			tribunus	*praetorian*	*proc ludi magni*		
					praef vehiculorum		

Table III.2. Careers of the *primipilares* of Dacia.

III.2.1. Particularities of the primpilate in Dacia

The situation of the *primi pili* and *primipilares* from Dacia largely corresponds with and confirms the above-mentioned characteristics. The number of attestations in the north-Danubian province is not very large, we know but 25 such characters. Regarding the situation as a whole, we can speak about a percentage of close to 20% *primipilares* from the total number of attested centurions, which would correspond to the real situation of that time, in fact portraying the real chances a centurion had to become *primus pilus*. Besides, I have considered only those who fulfill their military career in Dacia because those *primipilares* who arrive here as procurators, prefects or other such type of commanders belong to another class of officers and must be treated separately.[1006]

Obviously the legion *XIII Gemina* has the highest number of attestations in Dacia, with 18 characters. Eight, maybe nine of them are attested as *primipili*, while active or at the end of their one-year service. We know one of them from an official dedication, from the part of the whole legion, at the end of his mandate. Also two of them are attested on collective altars at Aquileia where they were probably detached with economic tasks at the middle of the 3rd century AD. A series are attested in different places of the Empire, probably those of origin: Philippi, Ancyra, Amastris, Ieso, Ravenna or Epetium. The rest are attested at Apulum, while performing private acts. One *primus pilus* of equestrian origin pays for a silver replica of the eagle and rebuilds the entrance to the *aedes* in the fortification from Apulum. The remaining are all *primipilares*. Only six of them are attested holding a superior post, all of them choosing to continue their military careers; three of them certainly become *praefecti castrorum* of the legion from Apulum. One of them enjoyed an iteration of the post of *praefectus castrorum* in the legion *IIII Flavia*. The remaining one might be considered commander *praefectus legionis*. It is the case of M. Aur. Veteranus,[1007] where we can understand things better because of the dating of the monument, under the rule of Gallienus – although not very accurately.

Evidence is a little scarce for the other Dacian legion, *V Macedonica*, with only five attestations of *primipilares*.[1008] Three of them are attested as *primipilares* and offer details about their subsequent careers. Two of them, both *praefecti legionis*, seem to have a different status. The first one, Donatus depends on the precise date of Gallienus's reform, although the dating of the monument is very tight, 255-256.[1009] This officer builds the temple of Azizos *bonus puer conservator*, in his official role, for the health of his family and for the legion from Potaissa. Perhaps this is not an official dedication like the one set up by the *primipilus* from Apulum,[1010] but in the context of Gallienus's reform he might be the highest-ranking officer in the area at that time. The second one, Optatus is considered, in principle, to be a *praefectus castrorum* dating back to Commodus. Finally, a *primipilus* from the legion *V Macedonica*, L. Artorius Castus has an exceptional subsequent career, commanding several vexillations as *praepositus* or *dux* and serving also as *praefectus castrorum* in a legion from Britain. After holding all of these command posts he is charged with the procuratorship of Liburnia, attached to Dalmatia – again a military post.[1011] Only two *primi pili* of *V Macedonica* are attested in service, but with no detaild connected to their career.[1012]

Finally, one of the most important characters originating in Dacia, the knight P. Aelius Marcellus does not seem to have any military connections with his home province: he obtains his primpilate in the legion *VII Gemina* and then is *praefectus castrorum* in the *VII Claudia* and *I Adiutrix*. But he is a knight from Apulum and was enlisted in the army directly in Rome, as centurion in the *castra peregrinorum,*[1013] an obvious advantage thanks to his social origin. In any case, he may lack an exceptional military career but he received tremendous civil honors. He is decurion and patron of the *colonia* Apulum and patron of three cities from Italy: Forum Flaminium, Fulginiae and Iguvium. This is a classic case of a *primipilaris* who minimized his military career – *praefectus castrorum* – but who has a very prolific civilian career, spanning across several provinces.

An unknown *primipilaris*, mentioned in an inscription in Rome[1014] can be added to this group: after the primpilate he holds command of a vexillation from the Dacian legions, as *dux leg(ionum) Dac(iae)*. He has a prolific equestrian career, is promoted to the procuratorship of Lusitania apparently only after completing the praetorian tribunate under Philippus Arabs. Subsequently he will become in Rome *procurator ludi magni* and *praefectus vehiculorum*. The extraordinary command from Dacia appears to be held before his primpilate.[1015] Because he does not mention the legions where he served as centurion and *primus pilus,* this might indicate one of the Dacian legions,[1016] although this cannot be proven.

Dacia, with its two legions, contributes greatly to the registry of *primipilares* found in the Empire. In regard of their origin, one can argue that the Italians dominate this registry. However the number of Italians overpasses those coming from Apulum only by two. And in any case, at this level, the career would have been very mobile and the hazard of discoveries can very well influence any statistic.

[1006] As they have been, see Piso 2013.
[1007] CIL III 1560=IDR III/1 54, Băile Herculane.
[1008] A *primus pilus* mentioned on a list of centurions discovered at Potaissa and so far unpublished can be added to this group. Information Prof. M. Bărbulescu.
[1009] CIL III 875, Potaissa.
[1010] CIL III 1092=IDR III/5 235.
[1011] CIL III 1919, 14224, Epetium, Salonae.
[1012] CIL III 910 and ICLP 10-11.
[1013] CIL III 7795=IDR III/5 442, CIL III 1181=IDR III/5 439, CIL III 1182=IDR III/5 441, Apulum and CIL XI 5215, 5216 Fulginiae.
[1014] CIL VI 1645=IDRE 19.
[1015] Dobson 1978, 304-305.
[1016] Cf. Petolescu, IDRE 19.

When it comes to the entry to the primipilate, we have in Dacia men from all the categories, except that of legionary soldiers, but in their case it is known that any mentions of the career before the centurionate is rare and supplemental. In this matter, the former praetorians, centurions or *principales*, overpass the former legionary centurions only by two, but we have to bear in mind that all of them must have passed through at least one legionary centurionate, before reaching the primipilate. Only one special case of a centurion of the *castra peregrinorum* is attested, and that is a local equestrian. Again we can deduce a slight advantage of the praetorians, but only if we consider the small pool of *primipilares* that make any mentions about their previous career. For all the others simple *primi pili* attested, any considerations about their origin or previous career is inaccurate.

Because most of them are observed while completing their primpilate or while holding the post of *praefectus castrorum*, we do not know *primipilares* who occupy municipal offices in Dacia with the exception of one who does so in other provinces as well. This situation might also be due to the fact that there are no veterans former *primipili* since, in fact, they never retire from military life. There is only a single *primus pilus*, coming from the legion *V Macedonica*, who has an equestrian career after his primpilate and this might reveal a lot about the real chances that a provincial *primus pilus* had to ascend procuratorships or prefectures. Most of them chose to end their military service as *praefecti castrorum*, thus marking an exceptional life and career.

It is certain that they played a vital role in the creation of Dacia's provincial elite, like other *primipilares* in other provinces. We will never know for sure how many of them left for other parts of the Empire to share their valuable expertise with other comrades in arms.

The Roman army of the Principate is an institution of huge proportions, spread over a very large territory and which had to function in perfect synchronization. Keeping in mind the technical limitations of such an endeavour and the changes that take place during a millennium of history, the Romans were always forced to improve this system, in every possible manner and, up to a certain point, they managed to keep up with this challenges.

The army served the interests of the moment for a long period of time, until it became a professional institution. As long as military service was only an obligation of the citizen, the Roman army existed with a severe shortage of human resources, but managed to raise up to the challenges of its times, namely the defense of Latium and Italy, of a relatively small area. Once the Roman state gradually expands the need to supplement the armed forces and thus to enlarge the recruitment pool was felt. These tendencies took shape only at the beginning of the 1st century BC, when the military profile of the Roman state was officially changed, but innovations of a practical nature were in use for a long time. During the whole Republican period the Romans had at their disposal exceptional military leaders but also some who were less inspired. Usually in the 'large-scale history' all responsibility falls on the shoulders of these characters, especially because information about other commanders who might influence the course of a war are missing or are small in number. The common point of all these commanders was the Roman army, the legion and the allies, later on auxiliaries, that they used as they thought best. The main problem of this system is that it does not practically reveal how it functioned, how the order from a commander, be it a consul or a *praetor*, reached the simple soldier, meaning from one man to almost 6,000.

If we apply the principles of the modern army it stands to reason that the Roman army must have had a very well articulated hierarchy, functioning on a well-defined structure, with approximately equal steps between each rank. That is why very many academics have searched for the 'missing links' from the Roman military hierarchy. The question that was asked – and it is still standing – is what happens – as far as hierarchy is concerned – between the only two officers whose domain of competence we know well: the military commander – consul, *praetor*, *legatus* and the commander of the smallest division of the legion – the centurion. The first rank to be interposed between the two was that of tribune. Even though it might seem logical from the hierarchical angle, we wonder why would the Romans task young men who just recently became adults with commanding large parts of the legion (1,ooo men)? Are they not rather 'students' in the command of the Roman army who train themselves for a military command that will come up eventually in their careers, earlier in the case of knights than in that of the senators? If this was so, then perhaps we should consider tribunes rather as officers from an important social background, mostly tasked with administrative duties, their main purpose during this period spent in the legion being to learn.

The next attempts descended on the hierarchical ladder. Starting from the premise that in the legion existed a very well organized centurion hierarchy, it was attempted to select a category among them – not necessarily at random but respecting the principles of organizing the legion in the battlefield – to command detachments of soldiers more numerous than a *centuria* or maniple. This idea is impossible to apply however, first of all for the Republican period. For this age information concerning a centurion hierarchy is completely lacking and even more so the yearly recruitment system rather indicated a hierarchical chaos at the lower level and not what the moderns desired, a centurion *primus inter pares*. Even the hierarchical role of the *primus pilus* was exaggerated for this period of time, because of his position in the line of battle and the length of service – one year. The question that remains however is at this point which centurionate post was held for more than one year?

How do we explain this 'vacuum' in the hierarchy of the Republican army? Probably simply because the legion is considered to be an indivisible military unit. We have no information about the legion's role of a 'modern' army, composed of several units able to function independently, despite its large size. The legion apparently needs only an organization along battle lines, three in total, or in groups of barracks in the military camps. We only know the rank of the legion's commander and that he has at his disposal other young 'apprentices' of the military command, finally that each contingent composed of 80-100 soldiers is under the command of a professional officer. To suddenly lower the level of command from 5,000-6,000 men to 60-100 seems unlikely and nonfunctional. From at least one point of view we are probably right to search for the 'missing links' in the legionary hierarchy, that of the circulation of information and orders. It is not possible to imagine a commander giving special orders to each of the 60 centurions before a battle or that a large number of them takes part in the war council. There has always been such a council except it contained only the legion's high ranking officers, but rather its structure depended on the commander's wishes. It is likely that these *amici* assist the commander, but not all of them are officers in the legion, thus they have no specific military attribute. It is just as likely that the commander's orders reach the centurions through the tribunes, but what we do not know and understand is how large the domain of competence of these tribunes is. This is true especially because they are of different ranks and we cannot simply divide the legion in 1,000 men units and place these purely theoretical divisions under each one's command.

We cannot ignore a very important aspect for the functioning of the legion, for as long as it existed, - its unity of action. On the battlefield it is only divided in battle lines, initially each one with its different role and equipment, each line being composed of 10 *manipuli*, the only division with its own officers. Vexillations appear too late and as a reaction to drastic changes in ancient warfare, therefore we cannot consider them divisions of the legion. This disproportionality is also strengthened by the apparent very good hierarchical organization of these subunits, the maniple and later on the *centuria*. The 'gap' in command does not disappear even after the professional reform of the Roman army although a new subdivision is created: the cohort, without its own commander. The result of this situation is that we are forced to consider all subdivisions without their own commanders as lacking any tactical role, having probably only an administrative one, and we are again left with only two tactical units: the legion and the *centuria*. Plus, we are motivated to accept the army's flawless practical skill, implying that everyone knew what he had to do and the legion operated as a whole. In fact sources that offer information about the 1st century BC seem to point out precisely such a system.

The Principate came as a natural evolution from this 1st century BC and institutionalized the supreme command of a superior officer over all army corps at his disposal – all of them in the case of Augustus. Obviously this command needed to be delegated, a situation when the first *princeps* called on the solutions achieved by his adoptive father, the legates, but maintaining the rigors of the Republican hierarchy. Curiously enough nothing changes inside the legion. Seniority and the primordial role of the centurion are officially recognized and consequently rewarded by the state. The most legitimate way to climb up the social ladder is established, from soldier to officer, from plebeian to knight, everything is part of the imperial policy of recruiting professional personnel. The three centuries of history that will follow will bring changes in the organization of the army but not in its functioning. The centurion stays the same during the whole of this period, only his access to superior ranks being gradually facilitated. After 250 years of applying the Augustan military policy, the commanding senators are removed, the army being exclusively filled with a professional staff fully trained at all its levels in the military art.

The common point of the whole period that was discussed is the centurion. His nature and function do not change from the beginnings of the Republican period, maybe even earlier, until the 4th century AD, for almost one millennium. Such constant must be due to an almost flawless lucrativeness of the system.

In the beginning the centurions are chosen to lead the 'handful' of men from the Republican legion, for which purpose they select a junior centurion, then each of them a second-in-command – *optio* and a standard bearer – *signifer*. This is the Roman army's hierarchical professional system that will remain in use during its entire existence with the exception of the period of the Later Empire. For the Republican period we have no information about a structured hierarchy of centurions, except the seniority of the centurion *primus pilus* in the form of his function and not necessarily rank, because he does not command a division of the legion greater than his maniple/*centuria*. Even the professionalization of the Roman army does not change the nature of the centurion. We begin to have more information about himself, his role in combat, even about his image, but there is nothing new in his area of expertise. We notice that the centurion has officially become an officer, fundamentally different from his soldiers but he remains one of the people. His role on the battlefield is prominent, for this period, the 1st century BC; now we have the most information about this aspect of his existence.

The system of the Principate superimposes that of Caesar's, including the problem of the centurionate. Except for the institutionalization of the possibility to advance to another social and military order that centurions with an exceptional career seem to enjoy, nothing appears to change concerning them. They remain the same commanders of the *centuria*, with the power of life and death over their soldiers and the same critical military role.

With regards to the origins of this category of officers we notice a normal evolution towards regionalization / provincialization during the Principate. This phenomenon cannot be placed in time only after two centuries AD, because individual attestations prove that it existed since the beginning of the imperial age. The provincialization of legionary and centurion recruitment evolves along with romanization of the provinces, meaning the establishing of Roman life and custom in the conquered provinces. It is thus normal that at the beginning of the Principate it is limited to Italy and the neighboring provinces, already part of the Roman state for hundreds of years. In the next two centuries the recruitment area will expand to include the whole Roman world in order to answer promptly and efficiently to the military needs of the state. As for maintaining and protecting the Roman-Italian element from the Empire's army through the distribution of officers, we cannot say that it was the official policy of any emperor, beyond the normal advantage enjoyed by the 'real' Romans from the cohorts in Rome. The legions however remain the same self-sufficient military units with the emperor simply appointing the commanding officer and confirming officially his policy of promotions. The added chance offered to centurions coming from Rome or upper social classes is only natural and logical since they fulfill all conditions needed for a long career: age, education, origin.

The access to the position of centurion remains unchanged while the post exists. Most centurions were former soldiers, promoted after a service of 10-15 years. These soldiers can come from the legion, most often promoted inside the same unit, or from the other categories of troops. In the beginning the exchange of officers was possible only between units recruited from citizens, meaning legions and praetorian cohorts, but the system will relax, meaning that entry to the centurionate will be available to soldiers from the auxiliary troops or the fleet as long as they have citizenship and the necessary training. By studying the careers of soldiers we can observe a series of posts or ranks from which promotion to centurion was granted, but again the system will evolve to a point when the centurionate will be available to every meritorious soldier or part of the centurion's attributions will be extended to soldiers. Part of the legionary centurions comes from the centurions serving in the cohorts in Rome, after transfer. These transfers are not part of a protectionist policy of the Empire towards the Roman-Italian part of the army, but the expression of a need to move around the officers and for the best of them to complete their primpilate in order to reach high positions. A third category for choosing centurions is the civilian body, provincial aristocrats, local elites or knights. Again we cannot imply that they are favored – compared to centurions from lower ranks – only because they have other means to advance in society and to mark these advancements for posterity. The base of the legion is represented by its legionaries and centurions, men who spend most of their adult lives in this environment and who obviously gain a professionalism that cannot be contested or equaled by other categories of centurions. The fact that most of them do not obtain higher positions is due to the length and nature of their work and because they did the same job, which was prestigious enough for most of them, for all their life.

The hierarchy of the centurions is by far the most difficult problem we have in understanding them. Their names and titles are already well known, all attempts to organize a hierarchy starting from here. Each centurion has a specific title, mentioning his cohort, thus the successive position in the line of battle, his placement in one of the three battle lines and his place in the old maniple. This title is not an invention of the Principate; it remains in use since the beginnings of the army organized on maniples, therefore we have no reason to give it any other meaning. A logical system for promoting centurions was attempted based on a very small number of inscriptions: it started with the tenth centurion from the third battle line and advanced towards the primpilate. Following a century of studies dedicated to this aspect it was not possible to establish a definite system for the hierarchy and promotion of centurions and this not only because of the very small number of such epigraphic attestations. On the contrary, all evolutions tend towards a much simpler theory, easy to apply in practice, that circulated parallel to the theory of a systemized hierarchy, both of them having numerous supporters. This theory claims that almost all the centurions from a legion are equal in rank. Their titles only serve to place them in the organization of the legion, when marching or on the battlefield and thus to make them easier to recognize. Also given that the military equipment and battle tactics evolved during the Principate the role of these men becomes even more important. With regards to their internal hierarchy it seems to lack meaning to try and standardize the existence of several ranks for the same officer, with the same command, over the same group of men. To this sense it was tried to differentiate between the attributions of centurions so that they could be placed on superior hierarchical steps. This search failed when it came to placing the centurion *pilus prior* at the command of the cohort in which he serves. There is no proof or evidence supporting such a situation besides our desire to create order even where it is missing or is not necessary. The situation is different concerning the centurions from the first cohort and the *primus pilus*. The first cohort of a legion had twice the number of soldiers and acted like an elite subunit, at least for an undetermined period of time. Its centurions commanded *centuriae* with double the number of soldiers and the *primus pilus* the whole maniple. This is sufficient reason for placing this category of centurions on a higher hierarchical step as compared to the other 54. Although we must not consider the doubling of the number of men from the first cohort as the establishment of an elite subunit but rather as a strengthening of the right flank of each battle line – a normal thing and probably an adaptation to the rigors of war at a certain time – still its centurions clearly occupy a different place in the hierarchy simply because they are called differently, *primi ordines*, as compared to the other, *centuriones*. The first three do not even use the full normal title – e.g. *I hastatus prior*; it is enough to mention to which battle line he belongs, *hastatus*, so that his position is known and in this case also his superior rank.

Given the present status of evidence and sources another conclusion concerning the hierarchy of the centurions would be at the very least poorly argued, especially because over 90% of epigraphic attestations that mentions centurions simply associate the two common terms, *centurio legionis*. The occasional attestation of the centurion's titles, besides those from the first cohort, can be explained away as a desire to underline other attributes of his rank, his position in the battle line probably having some signification connected with the armament, the training of recruits or battle tactics that is so far unclear to us.

The hierarchical equivalence of centurions can be observed from the multitude of functions that they have – besides commanding the *centuria*. None of these functions shows any preference for any category of centurions. Those who command the vexillations are not even *primi ordines*, their title having no value in this instance. The same is true for all the other 'uses' of the centurion; they are quite numerous as seen from the subchapter where this problem is discussed. Another argument in favor of equivalence is connected with money. Payment is the first symbol of rank in the Roman army and in the case of 54 centurions it is equal. Only the *primi ordines* have their payments increased – sometimes even twofold – again tilting the hierarchic scales in their favor. Finally, the social status and possibilities for social advancement are the same for all centurions and even the *primi ordines* cannot form a distinct group from this point of view.

Still, what can be said about the hierarchy and promotion system of centurions? First of all that probably most centurions served their whole careers in the same legion they enlisted, for a long period of time, because they were never veterans in the true meaning of the word. Movement from one post to another in the same legion was a normal result of the personnel policy. Some left the legion, disappeared; other replaced them, the need for new recruits – both soldiers and officers – being permanent. While in the army a centurion could be delegated with varied functions in the administration, police or justice. For the exercise of most of these functions he was sent away from his legion with or without the *centuria* under his command. These delegations could be permanent, especially for military reasons. Thus transfer is not a policy of personnel movement but a way to fill or supplement troops, according to local needs. The studies that follow the transfers of legionary centurions have shown that most of them follow the evolution of the military situation from different points of the Empire but only a few obvious cases can be considered a promotion. Sometimes necessities that are not of a military nature – e.g. technical expertize – lead to transfers of centurions.

To conclude, this 'complete officer' of the Roman army, for a millennium the best professional in the legion, was used by the Roman state for a variety of military and administrative tasks. Given his two characteristic attributes, training and authority, but also his modest social origin, he was the perfect agent to enforce imperial policy in the field and to delegate central provincial authority. The centurion – a privileged category of professional soldier – permanently enjoyed the opportunity for social advancement through the primpilate but still he continues to serve the emperor. The corps of approximately 2,000 centurions who were active in the army at a given time, combined with the *supernumerarii* and other such personnel, weigh heavily in the army and the provincial administration, this being the way to use them to their full potential.

The *primus pilus* is closely connected with the centurions. Considered the chief centurion of the legion, he was the highest-ranking professional officer in the legion since the time of the Republic. At that time he didn't differ from his colleagues as far as hierarchy is concerned as he too is a *centurio*; following the evolutions from the end of the Republic and the beginning of the Principate he will receive a privileged place and role, a special command as well as increased opportunities for promotion and social advancement. While his origins and promotion options are identical to those of the other centurions, the possibilities for continuing his career increase and become very promising.

First of all it is very likely that a centurion who obtains the primpilate holds this post for one year, a custom inherited from the time of the Republic. During this one year he is a privileged officer, a *primus inter pares*, he has the *aquila* and the legion's elite double *centuria in* his official care. Still, because he only holds this post for a short time, we should manifest reservation about giving him a critical tactical role in the legion or any extraordinary commands. The primpilate is considered – especially in peacetime – to be an honorific function, a crowning of a meritorious career and a path towards the equestrian order and another, higher, opportunities. In wartime the *primipili* are not only the best centurions and members in the war council but they also command important auxiliary troops or garrisons and are also authoritarian agents of peace and administration.
After the official end of the primpilate they receive one of the most renowned and sought after titles of the Imperial age: *primipilaris.* Once they finish their professional military service the emperor awards them the necessary census for joining the equestrian order that they join *de facto* from this moment. But in truth they never finish serving the emperor who helped them and they have two important paths opened up to them. The first means continuing the career in the legion, as *praefectus castrorum.* He is an officer with different duties – we do not know how military command fits in them – who is above the equestrian tribunes in the hierarchy. He is included in the war council and manages many logistical and administrative duties of the legion. For this he has at his disposal an *officium* made up of soldiers chosen from the legions and who are promoted or exempted from chores. It is the way in which a *primipilaris* with huge experience in the legion's business serves the army as a superior officer tasked with organizational duties and it is normal that this function be filled by a *primipilaris* with a legionary background. Usually it is the last post in one's career and ensures a high position and prestige for a centurion who was promoted from inside the legion.

His special place in Egypt's provincial army, where he probably was the *de facto* commander of the only legion from the province, the officer directly subordinated to *praefectus Aegypti*, and then his place in the Parthian legions, placed him in a good position for taking over all legions starting with the half of the 3rd century AD. This idea is supported by the contraction of his title from *praefectus castrorum legionis* to *praefectus legionis* and afterwards the equivalency with the title carried by equestrian commanders of the legion. This abbreviation only indicated that each legion was placed in a separate fortification and thus the need for a single prefect of the fortress. Also the extension of the situation of the Egyptian and *Parthicae* legions in the whole of the Empire does not place the *praefectus castrorum* in a command position because the prefects of those legions are *primipilares* with different careers who iterate their primpilate. Although we are not certain whether all *primipili II* belong in the legions, they hold the highest ranks after the removal of senators because they have completed the equestrian *militiae* in Rome and thus become the commanders of the legions.

This is the second career path available to *primipilares.* Normally the equestrian career is not a good choice for everybody; only the most prolific *primipilares* and those with the best origin and performance can enjoy such opportunities. They are given the opportunity to complete the equestrian *militiae* in Rome, as tribunes of the three, later four, types of groups stationed here (*vigiles*, *urbani*, praetorians and *equites singulares*) and to train for procuratorships. The equestrian *militiae* are always followed by the iteration of the primpilate, a post we do not completely understand and who may or may not have been an officer in the legion. While awaiting an opening and the opportunity of a procuratorian career, the *primipilares* remain active in a *numerus*, being delegated with important tasks by the emperor. But if they are given the possibility to follow a procuratorian career, the *primipilares*, because of their nature, tend to hold important military posts such as the prefectures of the praetorian fleets or command of the Egyptian legions, financial or presidial procuratorships or important military prefectures. There are few careers that evolve in this manner, especially among those coming from the legions but we can again observe the importance of this recruitment pool for the Roman state. Augustus himself organized this promotion path for centurions managing to gather the best in the group of professional officers used by all of his descendants.

Before becoming a centurion a simple legionary had to move through several posts or ranks that are placed between *immunes* and *principales*. Again literary and epigraphic sources do not present them with accuracy and suggest a very complicated career beneath the centurionate. The problem is much simpler however. From the point of view of the Roman army promotion meant an increase in payment thus below the centurionate we find only two military ranks corresponding to the *sesquiplicarius* and *duplicarius* payment levels. As previously mentioned the lower hierarchy of the legion is based on three ranks: *centurio*, *optio* and *signifer*. Academics tried to create some complicated promotion systems that occur in the careers of these soldiers, often ignoring the difference between rank and function. The principle mentioned above, hierarchy connected with payment, can help to organize most functions and posts that existed below the centurionate.

The first step in the career of a soldier was to be granted *immunis* status, meaning exemption from daily chores. This privileged status, already considered a promotion by the one who got it, is awarded only if that person has some special skill or ability to set him apart from his colleagues; this skill can be a technical one (*mensor, victimarius, architectus, medicus, capsarius, venator, horologiarius etc.*), it can be connected with education or administration (*librarius, exactus, exceptor, adiutor*), it might be that the soldier is selected in the guard detail of a superior officer (*singularis, strator, secutor*) and even abilities to operate special equipment or instruments are rewarded (*tubicen, cornicen, bucinator, haruspex, interprex*). Some of them are selected in the *officia* of lower-ranking officers such as tribunes or *praefecti castrorum*, as *beneficiarii*. Depending on how they fulfill their duties on these, very specialized, posts they can actually be promoted, meaning they get increased wages, becoming *principales*.

The first steps of the *principales* are the ranks and posts that are paid one and a half times more compared to a soldier's salary, the *sesquiplicarii*. We can include here most *principales* posts and ranks for which there are no attestation regarding promotion to the centurionate. The only ones that can be truly considered ranks, involving military activity in the *centuria*, are the *custos armorum*,[1017] if we do not include them in the *immunes* category, and the *tesserarius*. This does not mean that they the only *sesquiplicarii* in a legion, most such petty officers serving on different posts in the staffs of the superior officers or the governor (*frumentarius, quaestionarius, actarius, cornicularii* of the tribunes and of the *praefectus castrorum*). We are not certain whether or not to include in this group the category of specialized *optiones* (*optio valetudinarii, optio fabricae, optio custodiarum*), because they can also be considered *immunes* but their task of coordinating departments might mean a payment increase.

Finally the last step, of the *duplicarii,* is occupied by those posts and ranks that allow direct promotion to the centurionate. All standard bearers who serve in the legions (*aquilifer, imaginifer, signifer, vexillarius*) are situated on this rank because of their possibility to promote directly to centurion, despite attempts to include the *signifer* among the *sesquiplicarii. Optio* is a very important military rank, he is the deputy of the centurion and is the only petty officer holding some, if any, authority over the soldiers. Being marked as *optio spei* only ensures promotion to centurion at the first vacant place and is not a superior rank; the situation is similar to that of the *candidatus*. The category of posts receiving double payment and allow access to the centurionate includes most members of the *officium consularis* (*beneficiarius, speculator, commentariensis, cornicularius*) and the *cornicularius legati legionis*.

A separate category of petty officers who can be promoted directly to the centurionate are the centurions and decurions of the auxiliary units and the *evocati.* The first two categories can receive 4-6 times the salary of a legionary while the *evocati* might be those rarely attested *triplarii,* but they might earn even more.

To conclude, before becoming centurion a soldier's career included three ranks *immunis*, *sesquiplicarius* and *duplicarius* – both *principales*, and several posts. Once he reached the next pay level the soldier had several opportunities to hold a function that was paid just as well and some functions were held while waiting for subsequent promotions. We do not know what the length of service for such a function was, or for holding a rank of *principalis* but we know it was impossible to be demoted and that the soldier could not return to this previous normal pay level. This does not mean that all soldiers passed through the same posts and ranks but in most careers we notice evolutions that support this frame. The promoted soldier had to be trained in all aspects of military life, thus he served in both administrative and tactical posts. The centurionate was not guaranteed, many being discharged from a post of *duplicarius,* but it was a target everyone wished to reach. From the 3rd century onwards we notice a slight predominance of administrative posts perhaps due to the changes in the

[1017] In most of the specialty literature they are included in the *immunes* category.

soldiers' field of activity and their increased involvement in administration.

Even if we observe a 'pacification' of this type of soldiers and officers and that their specialization in the fields of administration and bureaucracy, this is a normal evolution. It starts when military conflicts are the main agenda of the Roman state and soldiers and officers are actual *viri militares* and deal with the main task of an army, war. It goes on to the periods of *pax romana*, when the army serves another role besides waging war, as agent charged with organizing and administrating the provinces and with spreading the Roman way of life in all the corners of the Empire. The extreme limit of this specialization will be crossed in the Later Empire when most of these officers will lose their original role and significance.

Epigraphic attestations from Dacia have helped to reach to the conclusions of this study just as much as those from other provinces. With regards to the lower hierarchy, Dacia has a large number of attestations, concerning both military ranks and the *officium consularis*. One of the best documented offices, the one from Dacia reveals possibilities for promotion to centurionate from a number of ranks considered inferior, such as the *speculator*. Also in Dacia we find mentioned a large group of *beneficiarii*, most of them of the governor, sent away in different *stationes* in the province.

With regards to the centurions and the *primi pili* the Roman army from Dacia has many such attestations, both inside and outside of the province. Most centurions attested in Dacia served their whole careers in one of the legions garrisoned in the province, part of them being transferred in several legions and provinces. Some of them are fulfilling special functions such as training the governor's guards. Approximately 20% of them, a normal percentage, reach the primpilate, most of them ending their careers with the post of *praefectus castrorum*. A small number followed a subsequent equestrian career commanding legions after Galliens or as *duces*; only two of them obtaining procuratorships. This does not mean Dacia is poor from this point of view but that if fits perfectly in the statistics and tendencies that can be notice for the whole of the Empire.

BIBLIOGRAPHY

Abreviations

AAESW - Abhandlungen des Archäologisch-Epigraphischen Seminars der Universität Wien, Vienna.
ACD – Acta Classica Universitatis Scientiarium Debrecenensis, Debrecen.
ActaMN – Acta Musei Napocensis, Cluj-Napoca.
AE – L'Année Épigraphique, Paris.
AHB –Ancient History Bulletin.
AJPh – American Journal of Philology, Baltimore.
AN – Aquileia Nostra, Aquileia.
AncSoc – Ancient Society, Leuven.
AnSt – Anatolian Studies, London.
ANRW – Aufstieg und Niedergang der römischen Welt, Berlin.
Arch – Archeologia, Warsaw.
BAR (IntS) – British Archaeological Reports (International Series).
BASP – Bulletin of the American Society of Papyrologists, New York.
BCTH – Bulletin Archéologique du Comté des Travaux Historiques, Paris.
BJb – Bonner Jahrbücher, Darmstadt.
CBFIR – Schallmayer, E. 1990 (ed.), Der römische Weihebezirk von Osterburken I. Corpus der griechischen und lateinischen Beneficiarier-Inschriften des Römischen Reiches, Stuttgart.
CE – Chronique d'Egypte, Bruxelles.
CIL – Corpus Inscriptionum Latinarum, Berlin.
CIG – Corpus Inscriptionum Graecarum, Berlin.
CJ – The Classical Journal, Athens.
CJb – Carnuntum Jahrbuch, Vienna.
ClA – Classical Antiquity, Berkeley.
CPh – Classical Philology, Chicago.
CQ – Classical Quarterly, Oxford.
CW – The Classical World, Pittsburg.
DNP – Der Neue Pauly. Encyclopädie der Antike, Stuttgart.
EN – Ephemeris Napocensis, Cluj-Napoca.
EpA – Epigraphica Anatolica, Bonn.
EpSt – Epigraphische Studien, Bonn.
HABES – Heidelberger Althistorische Beiträge und Epigraphische Studien, Stuttgart.
ICLP – Inscripţiile din castrul legionar de la Potaissa (Author: M. Bărbulescu).
IDR – Inscripţiile Daciei Romane (Authors: I.I. Russu, Gr. Florescu, C.C. Petolescu, I. Piso, V. Vollmann).
IDRE – Inscriptions externes concernant l'histoire de la Dacie (Author: C.C. Petolescu).
ILD – Inscripţii Latine din Dacia (Author: C.C. Petolescu).
JARCE - Journal of the American Research Center in Egypt, Boston.
JEA – Journal of Egyptian Archaeology, London.
JRA – Journal of Roman Archaeology, Portsmount-Rhode Island.
JRGZM – Jahrbuch des Römisch-Germanischen Zentralmuseums Mainz.
JRomMilSt – Journal of Roman Military Equipment Studies, Oxford.
JRS – Journal of Roman Studies, London.
KJb – Kölner Jahrbuch, Berlin.
LAq – Laureae Aquincenses, Budapest.
MAAR – Memoirs of the American Academy in Rome.
MAVORS – Roman Army Researches, Amsterdam.
MEFR - Mélanges de l' École Française de Rome, Paris.
PBSR – Papers of the British School at Rome, London.
RE – Paulys Real-Encyclopädie der classischen Altertumswissenschaft, Stuttgart.
REMA – Revue d' Etudes Militaires Anciennes, Paris.
RomO – Römisches Österreich, Vienna.
RSA - Rivista Storica dell'Antichità, Bologna.
SCIVA – Studii şi cercetări de istorie veche şi arheologie, Bucharest.
SJb – Saalburg-Jahrbuch, Berlin.
SPASCS - Supplementary Papers of the American School of Classical Studies.
TAPhA – Transactions and Proceedings of the American Philological Association, Cleveland.
TVind – Tabulae Vindolanenses (I, II, III, IV-1, IV-2, authors: A.K. Bowman, J.D. Thomas, R.S.O. Tomlin, J.N. Adams, J. Pearce).
WZ – Westdeutsche Zeitschrift für Geschichte und Kunst, Trier.
ZPE – Zeitschrift für Papyrologie und Epigraphik, Bonn.

Bibliography

Adams, J.N. 1999, *The Poets of Bu Njem: Language, Culture and the Centurionate*, JRS 89, 109-134.

Alexandrescu, Cristina-Georgeta 2005, *A Contribution on the Standards of the Roman Army*, Limes 19 Pécs, 147-151.

Alexandrescu, Cristina-Georgeta 2010, Blasmusiker und Standartenträger im römischen Heer. Untersuchungen zur Benennung, Funktion und Ikonographie, Cluj-Napoca.

Alföldy, G. 1979, Römische Sozialgeschichte^2, Wiesbaden.

Alföldy, G. 1988, The Social History of Rome, Baltimore.

Alföldy, G., Dobson, B., Eck, W. 2000, Kaiser, Heer und Gesellschaft in der Römischen Kaiserzeit, HABES 31, Stuttgart.

Allen, G.H. 1908, *The Advancement of Officers in the Roman Arny*, SPASCS II, 1-26.

Alston, R. 1994, *Roman Military Pay from Caesar to Diocletian*, JRS 84, 113-123.

Alston, R. 1995, Soldier and Society in Roman Egypt. A Social History, London/New York.

Ardevan, R. 1987, *Veterani şi decurioni municipali în Dacia romană*, Sargetia 20, 117-126.

Ardevan, R. 1991, *Beneficiarii în viaţa civilă a provinciei Dacia*, EN 1, 163-170.

Ardevan, R. 1994, *Die Beneficiarer im Zivilleben der Provinz Dakien*, Schallmayer 1994, 199-204.

Arnaud, P. 1995, *Les mensores des légions: mensores agrarii ou mensores frumentarii?*, Le Bohec 1995, 251-256.

Aubert, J.-J. 1995, *Policing the countryside: Soldiers and Civilians in Egyptian villages in the third and fourth centuries A.D.*, Le Bohec 1995, 257-266.

Austin, N.J.E., Rankov, N.B. 1995, *Exploratio*. Military and political intelligence in the Roman world from the Second Punic War to the battle of Adrianople, London/New York.

Baatz, D. 1989, *Kommandobereiche der Legionslegaten*, Germania 67, 169-178.

Baatz, D. 2000, *Vegetius' Legion and the Archaeological Facts*, Brewer 2000, 149-158.

Bagnall, R.S. 1977, *Army and Police in Roman Upper Egypt*, JARCE 14, 67-86.

Baillie Reynolds, P.K. 1923, *The Troops Quartered in the castra peregrinorum*, JRS 13, 168-189.

Baillie Reynolds, P.K., Ashby T. 1923, *The Castra Peregrinorum*, JRS 13, 152-167.

Baker, Patricia Anne 2004, Medical Care for the Roman Army on the Rhine, Danube and British Frontiers in the First, Second and Early Third Centuries AD, BAR (IS) 1286, Oxford.

Balla, L. 1968, *Vexillationes peregrinae im Heer des Septmius Severus*, EpSt 5, 143-148.

Bartoloni, R. 1996, *Alae, cohortes peditatae e cohortes equitatae: loro struttura e consistenza numerica*, RSA 26, 53-67.

Băluţă, C.L. 1978, *Un tubicen al legiunii a XIII-a Gemina la Apulum*, Apulum 16, 169-174.

Băluţă, C.L., Berciu, I. 1980, *Pedites şi equites singulares în Dacia. Materiale tegulare ştampilate*, Apulum, 18, 109-123.

Bărbulescu, M. 1987, Din istoria militară a Daciei romane. Legiunea *V Macedonica* şi castrul de la Potaissa, Cluj-Napoca.

Bărbulescu, M. 1997, Das Legionslager von Potaissa/Castrul legionar de la Potaissa, Zalău.

Bărbulescu, M. 2004, *Inscriptions votives pour les génies protecteurs dans le camp légionnaire de Potaissa*, Orbis Antiquus. Studia in Honorem Ioannis Pisonis, Cluj-Napoca, 375-379.

Bărbulescu, M. 2012, Inscripţiile din castrul legionar de la Potaissa – The Inscriptions of the legionary fortress at Potaissa, Bucureşti = ICLP.

Bell, M.J.V. 1965, *Tactical Reform in the Roman Republican Army*, Historia 14/4, 404-422.

Benea, Doina 1983, Din istoria militară a Moesiei Superior şi a Daciei. Legiunile a *VII*-a *Claudia* şi a *IIII*-a *Flavia*, Cluj-Napoca.

Bennett, J. 2006, *New Evidence from Ankara for the collegia veteranorum and the alabata decursio: In Memoriam J.C. Mann*, AnSt 56, 95-101.

Bennett, J. 2007, *Two New centurions of the legio IIII Scythica*, Latomus 66/2, 404-413.

Bérard, F. 1995, *La cohorte urbaine de Lyon: une unité à part dans la Rangordnung?*, Le Bohec 1995, 373-382.

Berciu, I., Popa, A. 1964, *Exceptores consularis in Dacia*, Latomus 23, 302-310.

Best Jr., E.E 1966, *The Literate Roman Soldier*, CJ 62/3, 122-127.

Birley, A.R. 1979, The People of Roman Britain, London.

Birley, A.R. 2005, The Roman Government of Britain, Oxford.

Birley, E. 1941, *The Origins of Legionary Centurions*, LAq 11, 47-62 (MAVORS IV, 189-205).

Birley, E. 1964, *Promotions and Transfers in the Roman Army II: The Centurionate*, CJb 21-33 (MAVORS IV, 206-220).

Birley, E. 1966, *Alae and cohortes milliariae*, Corolla Memoriae Erich Swoboda Dedicata, Graz-Köln, 54-67.

Birley, E. 1978, *Alae named after their commanders*, AncSoc 9, 257-273.

Birley, E. 1978a, *The Religion of the Roman Army: 1895-1977*, ANRW II.16.2, 1506-1541.

Birley, E. 1981, *Evocati Aug.: A Review*, ZPE 43, 25-29.

Birley, E. 1984, *A Centurion of Leg. VI Victrix and his Wife*, ZPE 57, 230-232.

Birley, E. 1989, *Some legionary centurions*, ZPE 79, 114-128.

Bishop, M.C., Coulston J.C.N 2006, Roman Military Equipment. From the Punic Wars to the Fall of Rome^2, Oxford.

Blommaert, D.C. 1989, *Tribunus equitum singularium? Two Fragmentary Inscriptions Concerning Egyptian Legionary Prefects*, ZPE 79, 297-307.

Boren, H.C. 1983, *Studies relating to the stipendium militum*, Historia 32/4, 427-460.

Bowman, A.K. 1994, Life and Letters on the Roman Frontier, London.
Brand, C.E. 1968, Roman Military Law, Austin/London.
Breeze, D.J. 1969, *The Organization of the Legion: the First Cohort and the Equites Legionis*, JRS 59, 50-55.
Breeze, D.J. 1971, *Pay Grades and Ranks Below the Centurionate*, JRS 61, 130-135.
Breeze, D.J. 1974, *The Organization of the Career Strucuture of the immunes and principales of the Roman Army*, BJb 174, 245-292.
Breeze, D.J. 1976, *A Note on the Use of the Titles optio and magister below the Centurionate during the Principate*, Britanna 7, 127-133.
Breeze, D.J. 1976a, *The Ownership of Arms in the Roman Army*, Britannia 7, 93-95.
Breeze, D.J. 1993, *Cavalry on Frontiers*, Roman Officers and Frontiers (MAVORS X), Stuttgart, 288-297.
Brewer, R.J. (ed.) 2000, Roman Fortresses and their Legions, London.
Brizzi, G. 1995, *La gerarchia militare in età repubblicana*, Le Bohec 1995, 15-22.
Brown, R. 2004, *"Virtus consili expers": An Interpretation of the Centurion's Contest in Caesar, De bello Gallico 5.44*, Hermes 132/3, 292-308.
Brunt, P.A. 1950, *Pay and Superannuation in the Roman Army*, PBSR 18, 50-71.
Brunt, P.A. 1983, *Princeps and Equites*, JRS 73, 42-75.
Bruun, C. 1988, *Caligatus, tubicen, optio carceris, and the Centurions' Positions; Some Remarks on an Inscription in ZPE 71 (1988)*, Arctos 22, 23-40.
Buora, M. 1994, *Beneficiarii in Aquileia*, AN 65, 65-80.
Campbell, B. 1984, The Emperor and the Roman Army (31 BC-AD 235), Oxford.
Campbell, B. 1987, *Teach Yourself How to be a General*, JRS 77, 13-29.
Campbell, B. 1994, The Roman Army. A Source Book, London/New York.
Campbell, B. 2001, *Primipilus*, DNP 10, 326-327.
Campbell, B. 2002, War and Society in Imperial Rome (31 BC-AD 284), London/New York.
Campbell, B. 2004, Greek and Roman Military Writers. Selected Readings, London/New York.
Castillo, Carmen 1995, *Primipilares* 1978-1992, Le Bohec 1995, 211-220.
Cheesman, G.L. 1914, The Auxilia of the Roman Imperial Army, Oxford.
Christol, M. 1977, *La carrière de Traianus Mucianus et l'origine de protectores*, Chiron 7, 393-408.
Christol, M., Demougin, S. 1986, *Un préfet de camp de la légion troisième Cyrenaïque*, ZPE 64, 195-199.
Christol, M. 1994, *Sur quelques centurions de la légion III Augusta*, ZPE 103, 101-107.
Clauss, M. 1973, Untersuchungen zu den principales des römischen Heeres von Augustus bis Diokletian. *Cornicularii, speculatores, frumentarii, Diss*. Bochum.
Clauss, M. 1975, *Der canalicularius*, AncSoc 6, 251-256.
Clauss, M. 1980, *Frumentarius Augusti*, Epigraphica 42, 131-134.
Cupcea, G., Marcu, F. 2006, *The Size and Organization of the Roman Army and the Case of Dacia under Trajan*, Dacia NS 50, 175-194.
Cupcea, G. 2008, *Speculatores in Dacia. Missions and Careers*, ActaMN 43-44/I, 263-279.
Cupcea, G. 2009, *The missions of the soldiers in the limes provinces. Frumentarii in Dacia*, Near and Beyond the Roman Frontiers (ed. O Ţentea, I.C. Opriş), Bucureşti, 305-314.
Cupcea, G. 2012, *Officium consularis. The Evidence of Dacia*, Transylvanian Review XXI Suppl. 3, 243-254.
Cupcea, G. 2012a, *Veteran Settlement and Colonia Ulpia Traiana Sarmizegetusa*, Scripta Classica. Radu Ardevan sexagenario dedicata (eds. I. Piso, V. Rusu-Bolindeţ, R. Varga, S. Mustaţă, E. Beu-Dachin, L. Ruscu), Cluj-Napoca, 247-255.
Cupcea, G. 2013, *Centurions of the IIII Flavia Legion in Dacia*, Arheovest 1. *In memoriam Liviu Măruia*, Timişoara, 343-356.
Cupcea, G. 2014, forthcoming, *A 'horologiarius' from Apulum and Time-Keeping in the Roman Army*.
Cupcea, G. 2015, forthcoming, *Careers in the Guard of the Dacia Governor*, Dacia NS 58, forthcoming.
Dabrowa, E. 1993, *Legio X Fretensis*. A Prosopographical Study of its Officers (I-III c. AD), Stuttgart.
Davies, R.W. 1969, *The medici of the Roman Armed Forces*, EpSt 8, 83-99.
Davies, R.W. 1970, *The Roman Military Medical Service*, SJb 27, 84-104.
Davies, R.W. 1971, *Cohortes Equitatae*, Historia 20/5-6, 751-763.
Davies, R.W. 1972, *Some more Military medici*, EpSt 9, 1-11.
Davies, R.W. 1976, *Centurions and Decurions of cohors XX Palmyrenorum*, ZPE 20, 253-275.
Davies, R.W. 1976a, *Singulares and Roman Britain*, Britannia 7, 134-144.
Davies, R. 1989, Service in the Roman Army, Edinburgh.
Davison, D.P. 1989, The Barracks of the Roman Army from the 1st to 3rd Centuries AD, BAR (IS) 472.
De Blois, L. 1987, The Roman Army and Politics in the First Century Before Christ, Amsterdam.
De Blois, L. 2007, *Army and General in the Late Roman Republic*, Erdkamp 2007, 164-180.
Dessau, H. 1910, *Die Herkunft der Offiziere und Beamten Römischen Kaiserreichs. Während der Ersten Zwei Jahrh. Seines Bestehens*, Hermes 45/1, 1-26 and 45/4, 615-617.
Dessau, H. 1912, *British centurions*, JRS 2, 21-24.
Devijver, H. 1970, *Suétone, Claude, 25, et les milices équestres*, AncSoc 1, 69-81.
Devijver, H. 1974, *The Roman Army in Egypt (with a special reference to the militiae equestres)*, ANRW II.1, 452-259.
Devijver, H. 1974a, *The Age of Equestrian Officers*, MAVORS VI, 73-140.
Devijver, H. 1976-1977, *Prosopographia militiarum equestrium qui fuerunt ab Augusto ad Gallienum* I-III şi Suppl. I-II (1987, 1993), Leuven.
Devijver, H. 1989, *L'Egypte et l'histoire de l'armée romaine*, MAVORS IX, 22-39.

Devijver, H. 1990, *Bears and Bizons ... and the Roman Army*, Kleio 20, 87-97, in English in MAVORS IX, 140-147.

Devijver, H. 1995, *Les milices équestres et la hiérarchie militaire*, Le Bohec 1995, 175-191.

Dise, R.L. 1995, *A Reassessment of the Functions of beneficiarii consularis*, AHB 9/2, 72-85.

Dise, R.L. 1996, *The beneficiarii procuratoris of Celeia and the Development of the statio Network*, ZPE 113, 286-292.

Dise, R.L. 1997, *Trajan, the Antonines, and the Governor's Staff*, ZPE 116, 273-283.

Dise, R.L. 1997a, *Variation in Roman Administrative Practice: The Assignements of beneficiarii consularis*, ZPE 116, 284-299.

Dixon, Karen R., Southern, Pat 1992, The Roman Cavalry. From the First to the Third Century A.D., London.

Dobó, A. 1968, Die Verwaltung der römischen Provinz Pannonien von Augustus bis Diocletianus, Budapest.

Dobó, A. 1978, *L'officium consularis en Dacie*, ACD 14, 57-64.

Dobson, B., Breeze, D.J. 1969, *The Rome Cohorts and the Legionary Centuionate*, EpSt 8, 100-124.

Dobson, B. 1970, *The Centurionate and Social Mobility during the Principate*, Recherches sur les structures sociales dans l`Antiquité classique (ed. Cl. Nicolet), Paris, 99-115.

Dobson, B. 1972, *Legionary Centurion or Equestrian Officer? A Comparison of Pay and Prospects*, AncSoc 3, 193-207.

Dobson, B., Mann, J.C. 1973, *The Roman Army in Britain and Britons in the Roman Army*, Britannia 4, 191-205.

Dobson, B., 1974, *The significance of the centurion and primipilaris in the Roman army and administration*, ANRW II-1, 392-434.

Dobson, B. 1978, Die *Primipilares*, Bonn.

Dobson, B. 1979, *The `Rangordnung` of the Roman Army*, Actes du VIIe Congrès International d`Épigraphie Grecque et Latine (ed. D.M. Pippidi), Bucharest-Paris, 191-204.

Dobson, B. 1982, *Praefectus castrorum Aegypti – a reconsideration*, CE 57/114, 322-337.

Dobson, B. 1986, *The Roman Army: Wartime or Peacetime Army?*, Heer und Integrationspolitik 2, Köln, 10-25.

Dobson, B. 1995, *The Bibliography of Rangordnung*, Le Bohec 1995, 41-46.

Dobson, B. 2000, *The primipilares in Army and Society*, Alföldy, Dobson, Eck 2000, 139-152.

Dolata, J. 1995, *Promotio militaris – die Beförderung eines ritterlichen Offiziers*, Klio 77, 255-265.

Domaszewski, A. von 1885, *Die Fahnen im römischen Heere*, AAESW5, 1-80.

Domaszewski, A. von 1895, *Die Religion des römischen Heeres*, WZ 14-1, 1-129.

Domaszewski, A. von 1908, Die Rangordnung des römischen Heeres, Bonn.

Domaszewski, A. von-Dobson, B. 1967, Die Rangordnung des römischen Heeres², (ed. B. Dobson) Köln-Graz.

Dubuisson, M. 1996, *Renseignement, espionnage et services secrets dans l'armée romaine*, Ktema 21, 305-313.

Ducos, Michèle 1995, *La hiérachie militaire dans les sources littéraires*, Le Bohec 1995, 47-52.

Durry, M. 1968, Les cohortes prétoriennes², Paris.

Duval, N., Lancel, S., Le Bohec, Y. 1984, *Études sur la Garnison de Carthage*, BCTH 15-16 B, 33-89.

Eck, W., Wolff, H. (ed.) 1986, Heer und Integrationspolitik. Die römischen Militärdiplome als historische Quelle, Köln/Wien.

Eibl, Kordula 1994, *Gibt es eine spezifische Ausrüstung der Beneficiarier*, Schallmayer 1994, 273-298.

Epplett, C. 2001, *The Capture of Animals by the Roman Military*, Greece & Rome II.48/2, 210-222.

Erdkamp, P. 2007 (ed.), A Companion to the Roman Army, Oxford.

Evans, Edith 1994, *Military Architects and Building Design in Roman Britain*, Britannia 25, 143-164.

Faure, P. 2008, *Les symboles des centuries légionnaires*, ZPE 165, 293-303.

Feig Vishnia, Rachel 2002, *The Shadow Army – The Lixae and the Roman Legions*, ZPE 139, 265-272.

Fellmann, R. 1995, *L'inscription d'un "optio princeps" au temple de Ba'alshamîn à Palmyre*, Le Bohec 1995, 239-240.

Feugère, M. 1995, *L'équipment des officiers dans l'armée Romaine*, Le Bohec 1995, 113-126.

Fields, N. 2007, The Roman Army of the Punic Wars 264-146 BC, Battle Orders 27, Oxford.

Fields, N. 2008, The Roman Army: The Civil Wars 88-31 BC, Battle Orders 34, Oxford.

Fields, N. 2009, The Roman Army of the Principate 27 BC-AD 117, Battle Orders 37, Oxford.

Fink, R.O. 1947, *The Cohors XX Palmyrenorum, a Cohors Equitata Milliaria*, TAPhA 78, 159-170.

Fink, R.O. 1953, *Centuria Rufi, Centuria Rufiana and the Ranking of Centuries*, TAPhA 84, 210-215.

Fink, R.O. 1971, Roman Military Records on Papyrus, Michigan.

Forni, G. 1953, Il recrutamento delle legioni da Augusto a Diocleziano, Milano/Roma.

Forni, G. 1974, *Estrazione etnica e sociale dei soldati delle legioni nei primi tre secoli dell'impero*, ANRW II.1, 339-391.

Freis, H. 1962, Die *cohortes urbanae*, Epigraphische Studien 5, Merchweiler.

Frere, S.S. 1980, *Hyginus and the First Cohort*, Britannia 11, 51-60.

Fuhrmann, C. 2012, Policing the Roman Empire. Soldiers, Administration and Public Order, Oxford.

Gilliam, J.F. 1940, *The ordinarii and ordinati of the Roman Army*, TAPhA 71, 127-148.

Gilliam, J.F. 1946, *Milites caligati*, TAPhA 77, 183-191.

Gilliam, J.F. 1956, *The Veterans and praefectus castrorum of the II Traiana AD 157*, AJPh 77, 359-375.

Gilliam, J.F. 1958, *The Appointment of Auxiliary Centurions* (P. *Mich* 164), TAPhA 88, 155-168 = MAVORS II, 191-207.

Gilliam, J.F. 1976, *Canaliclarius and kananiklarios (P. Oxy XL 2925)*, BASP 13, 49-52.

Gilliver, Kate, 2007, *Display in Roman Warfare: The Appearance of Armies and Individuals on the Battlefield*, War in History 14/1, 1-21.
Goldsworthy, A.K. 1996, The Roman Army at War. 100 BC-AD 200, Oxford.
Goldsworthy, A. 2005, Essential Histories. Caesar's Civil War 49-44 BC, New York/London.
Goldsworthy, A.K. 2008, Totul despre armata romană, Bucureşti.
Gómez, Esther Checa, Storch de Garcia, J.J. 2009, *Intérpretes militaires*, Limes 20 Léon, 1039-1050.
González, A.V. 1970, *Legio VII Gemina*, Léon.
Gudea, N. 1976, *Legio VII Gemina în Dacia*, SCIVA 27/1, 109-113.
Haensch, R. 1995 – *A comentariis und comentariensis: Geschichte und Aufgaben eines Amtes im Spiegel seiner Titulaturen*, Le Bohec 1995, 267-284.
Haensch, R. 2001, *Inschriften un Bevölkerunsgeschichte Niedergermaniens. Zu den soldaten der Legiones I Minervia und XXX Ulpia Victrix*, KJb. 34, 89-134.
Haensch, R. 2001a, *Veteranus ex beneficiario consularis, equestris militiae petitor. Mögliche Gründe für einen Aussergewöhnlichen Aufstieg*, KJb. 34, 135-139.
Harmand, J. 1967, L'armée et le soldat à Rome de 107 à 50 avant notre ère, Paris.
Haynes, I. 2001, *The Impact of Auxiliary Recruitment on Provincial Societies from Augustus to Caracalla*, Administration, Prosopography and Appointment Policies in the Roman Empire (ed. L. De Blois), Amsterdam, 62-83.
Herz, P. 1975, *Honos aquilae*, ZPE 17, 181-197.
Herz, P. 1999, *Der centurio supernumerarius und die annona militaris*, Laverna 10, 165-184.
Herz, P. 2007, *Finances and Costs of the Roman Army*, Erdkamp 2007, 306-322.
Hirschfeld, O. 1891, Die Sichercheitspolizei im römischen Kaiserreich, Berlin.
Hirschfeld, O. 1893, Die *agentes in rebus*, Berlin.
Hirschfeld, O. 1963, Die kaiserlichen Verwaltungsbeamten bis auf Diocletian[3], Berlin.
Hirt, A.M. 2010, Imperial Mines and Quarries in the Roman World. Organizational Aspects 27 BC-AD 235, Oxford.
Hoffmann, Brigitta 1995, *The Quarters of Legionary Centurions of the Principate*, Britannia 26, 107-151.
Isaac, B. 1990, The Limits of the Empire. The Roman Army in the East, Oxford.
Isaac, B. 1995, *Hierarchy and Command-Structure in the Roman Army*, Le Bohec 1995, 23-31.
Isac, D. 1994, *Vicus Samum – eine statio der Beneficiarier an der nördlichen Grenze Dakiens*, Schallmayer 1994, 205-218.
Isac, D. 1997, Castrele de cohortă şi ală de la Gilău, Zalău.
Isac, D. 2003, Castrul roman de la Samum-Căşeiu, Cluj-Napoca.
Junkelmann, M. 1986, Die Legionen des Augustus, Mainz.
Kehne, P. 2007, *War- and Peacetime Logistics: Supplying Imperial Armies in East and West*, Erdkamp 2007, 323-338.
Kennedy, D.L. 1983, *Milliary cohorts: The Evidence of Josephus, BJ, III.4.2 (67) and of Epigraphy*, ZPE 50, 253-263.
Kennedy, D.L. 1985, *The construction of a vexillation from the army of Syria and the origin of alae milliariae*, ZPE 61, 181-185.
Kennedy, D. 1985a, *The Composition of a Military Work Party in Roman Egypt (ILS 2483: Coptos)*, JEA 71, 156-160.
Keppie, L. 1984, The Making of the Roman Army. From Republic to Empire, London.
Keppie, L. 1984a, *Colonisation and Veteran Settlement in Italy in the First Century AD*, PBSR 52, 77-114.
Keppie, L. 1991, *A centurion of legio Martia at Padova?*, JRomMilSt 2, 115-121.
Keppie, L. 1993, *Vexilla veteranorum*, PBSR 41, 8-16.
Keppie, L. 1996, *The Praetorian Guard before Sejanus*, Athenaeum 84, 101-124.
Keyes, K.W. 1915, The Rise of the *Equites* in the 3rd Century of the Roman Empire, Princeton.
Kleiser, C. 2003, *Miniatur-Benefiziarierlanzenspitzen aus dem Raum Carnuntum*, RomO 26, 101-108.
Kolendo, J. 1980, *Le rôle du primus pilus dans la vie religieuse de la légion*, Arch 31, 49-60.
Künzl, E. 2008, Unter den goldenen Adlern. Der Waffenschmuck der römischen Imperiums, Regensburg/Mainz.
Lammert, F. 1954, *Primipilus*, RE XXII 2, 1974-1976.
Landels, J.G. 1999, Music in Ancient Greece and Rome, London/New York.
Le Bohec, Y. 1989, La troisième légion Auguste, Paris.
Le Bohec, Y. 1994, The Imperial Roman Army, London.
Le Bohec, Y. 1995 (ed.), La hiérarchie (*Rangordnung*) de l'armée Romaine sous le Haut- Empire (Actes du Congrès de Lyon, 15-18 septembre 1994), Paris.
Le Bohec, Y. 1995a, *Pour servir à l'étude de la hiérachie dans l'armée romaine du Haut-Empire*, Le Bohec 1995, 11-14.
Le Bohec, Y., Wolff, Catherine 2000 (ed.), Les légions de Rome sous le Haut-Empire (Actes du Congrès de Lyon, 17-19 septembre 1998), Lyon.
Le Bohec, Y. 2012, *Les petitores de l'armée romaine*, Le métier de soldat dans le monde romain (ed. C. Wolff), Lyon, 2012, 513-524.
Le Roux, P. 1982, L'armée romaine et l'organisation des provinces Ibériques, Paris.
Le Roux 1992, *L'armée Romaine sous les Sévères*, ZPE 94, 261-268.
Lendon, J.E. 2006, *Contubernalis, commanipularis and commilito in Roman Soldiers' Epigraphy: Drawing the Distinction*, ZPE 157, 270-276.
Lendon, J.E. 2005, Soldiers & Ghosts. A History of Battle in Classical Antiquity, New Haven/London.
Letta, C. 1978, *Le imagines Caesarum di un praefectus castrorum Aegypti e l'XI coorte pretoria*, Athenaeum 56, 3-19.
Lopuszański, G. 1938, *La transformation du corps des officiers supérieurs dans l'armée romaine du Ier au IIIe siècle après J.-C.*, MEFR 55, 131-183.
Lőrincz, B. 2001, Die römischen Hilfstruppen in Pannonien während der Prinzipatszeit. Teil I: Die Inschriften, Wien.

Luceroni, Maria 2001, Gli stationarii in eta Imperiale, Roma.

Luttwak, E.N. 1979 - The Grand Strategy of the Roman Empire from the I[st] century AD to the IIIrd, Baltimore-London.

MacKenzie, D.C. 1983, *Pay Differentials in the Early Empire*, Classical World 76/5, 267-273.

MacMullen, R. 1984, *The Legion as a Society*, Historia 33/4, 440-456.

Malone, S.J. 2006, *Legio XX Valeria Victrix*. Prosopography, archaeology and history, BAR (IS) 1491, Oxford.

Mann, J.C. 1954, *A Note on the numeri*, Hermes 82, 501-506.

Mann, J.C. 1983, Legionary Recruitment and Veteran Settlement during the Principate, London.

Mann, J.C. 1983a, *Trecenarius*, ZPE 52, 136-140.

Mann, J.C. 1988, *The Organization of the frumentarii. The Castra Peregrina and the peregrini*, ZPE 74, 148-150.

Mann, J.C. 1993, *Cavalry Promotion at Kirkby Thore*, Britannia 24, 261-264.

Mann, J.C. 1997, *Roman Legionary Centurial Symbols*, ZPE 115, 295-298.

Marquardt, J. 1876, Römische Staatsverwaltung II, Leipzig.

Maxfield, Valerie 1981, The Military Decorations of the Roman Army, London.

McDermott, W.C. 1970, *Milites Gregarii*, Greece & Rome 2nd Series 17/2, 184-196.

Mench Jr., F.C. 1968, The *cohortes urbanae* of the Imperial Rome. An Epigraphic Study, Michigan.

Merkelbach, R. 1998, *Centurio regionarius (zu Plinius, Epist. X 77/8)*, EpA 30, 114.

Milner, N.P. 1993, Vegetius: Epitome of Military Science, Liverpool.

Mircović, Miroslava 1994, *Beneficiarii consularis in Sirmium*, Schallmayer 1994, 193-198 and Chiron 24, 345-404.

Mircović, Miroslava 2004, *The Roster of the VII Claudia Legion*, ZPE 146, 211-220.

Moga, V. 1975, *Prefecţi ai castrului legiunii XIII Gemina la Apulum*, Apulum 13, 651-656.

Moga, V. 1985, Din istoria militară a Daciei romane. Legiunea *XIII Gemina*, Cluj-Napoca.

Moga, V., Drâmbărean, M. 2007, *Note epigrafice apulense*, Apulum 44, 175-179.

Moga, V., Piso I., Drâmbăreanu, M. 2008, *L'aigle de la legio XIII Gemina*, ActaMN 43-44/I, 177-184.

Morizot, P. 1995, *Ex tubicine principe, ex tubicine principis, ou ex tubicine, princeps...*, Le Bohec 1995, 241-244.

Mosser, M. 2003, Die Steindenkmäler der *legio XV Apollinaris*, Wien.

Mócsy, A. 1996, *Das Lustrum primipili und die Annona Militaris*, Germania 44, 312-326.

Nelis-Clément, Jocelyne 2000, Les *beneficiarii*: militaires et administrateurs au service de l'Empire (Ier s. a.C. – VIe s. p.C.), Bordeaux.

Nemeti, S. 2007 – *Note epigrafice. CIL III 7688*, Dacia Felix. Studia Michaeli Bărbulescu oblata, Cluj-Napoca, 229-234.

Odahl, C.M. 2004, Constantine and the Christian Empire, London/New York.

Oorthuijs, J. 2007, *Marines and Mariners in the Roman Imperial Fleets*, Impact of Empire 6 (ed. De Blois, Lo Cascio), Leiden/Boston, 169-180.

Opreanu, C. 1994, *Misiunile beneficiarilor consulari pe limes-ul de nord al Daciei în secolul al III-lea*, ActaMN 31/I, 69-77.

Opreanu, C. 1998, *Castrul legiunii I Adiutrix de la Apulum în timpul lui Traian. Consecinţe asupra urbanismului roman la Apulum*, Apulum 35, 121-134.

Osier, J. 1977, *The Emergence of Third-Century Equestrian Military Commanders*, Latomus 36/2, 674-687.

Ott, J. 1994, *Überlegungen zur Stellung der Beneficiarier in der Rangordnung des römischen Heeres*, Osterburken II, 233-250.

Ott, J. 1995, Die Beneficiarier, Stuttgart.

Ott, J. 1995a, *Die Mechanismen bei der Beförderung von Beneficiariern der Statthalter*, Le Bohec 1995, 285-290.

Parker, H.M.D. 1926, *A Note on the Promotion of the Centurions*, JRS 16, 45-52.

Parker, H.M.D. 1928, The Roman Legions, Oxford.

Parker, H.M.D. 1932, *The antiqua legio of Vegetius*, CQ 26/3-4, 137-149.

Passerini, A. 1939, Le coorti pretorie, Roma.

Patterson, J. 1993, *Military Organization and Social Change in the Later Roman Republic*, War and Society in the Roman World (eds. J. Rich, G. Shipley), London, 92-112.

Pavkovič, M.F. 1994 - *Singulares legati legionis: Guards of a Legionary Legate or a Provincial Governor?*, ZPE 103, 223-228.

Peretz, D. 2006, *The Roman Interpreter and His Diplomatic and Military Roles*, Historia 55/4, 451-470.

Petolescu, C.C. 1995, *Q. Trebellius maximus ex trecenario*, Le Bohec 1995, 245-248.

Petolescu, C.C. 2002 - *Auxilia Daciae*, Bucureşti.

Petolescu, C.C. 2005, *Cariera centurionului M. Calventius Viator*, Pontica 37-38, 195-198.

Petolescu, C.C. 2005a, Inscripţii latine din Dacia, Bucureşti = ILD.

Pflaum, H.-G. 1960-1961, Les carrières procuratoriennes équestres sous le Haut-Empire romain I-III, Paris.

Pflaum, H.G. 1976, *Zur Reform des Kaisers Gallienus*, Historia 25/1, 109-117.

Phang, Sara Elise 2008, Roman Military Service. Ideologies of Discipline in the Late Republic and Early Principate, Cambridge.

Piret, F. 1995, *Les dévotions des centurion de l'armée romaine du Rhin au miroir de la documentation épigraphique*, Le Bohec 1995, 231-237.

Piso, I. 1980, *Epigraphica (XI)*, Potaissa. Studii şi comunicări II, 123-131.

Piso, I., Benea, D. 1984 - *Das Militärdiplom von Drobeta*, ZPE 56, 263-295.

Piso, I. 1993, *Fasti provinciae Daciae* I. Die senatorichen Amtsträger, Antiquitas 43, Bonn.

Piso, I. 1994 - *Ein unechter Beneficiarius aus Apulum*, ZPE 103, 207-208.

Piso, I., Benea, D. 1999 – *Epigraphica Tibiscensia*, ActaMN 36/1, 91-107.
Piso, I. 2000, *Les légions dans la province de Dacie*, Le Bohec, Wolff 2000, 205-227.
Piso, I. 2005, An der Nordgrenze des Römischen Reiches. Ausgewählte Studien (1972-2003), HABES 41, Stuttgart.
Piso, I. 2008, *Les débuts de la province de Dacie*, Die römischen Provinzen. Begriff und Gründung (ed. I. Piso), Cluj-Napoca, 297-332.
Piso, I. *et al* 2012, Der Marmor im römischen Dakien (H.W. Müller, B. Schwaighofer, I. Piso, M. Benea) Cluj-Napoca.
Piso, I. 2013, *Fasti provinciae Daciae* II. Die ritterlichen Amtsträger, Antiquitas 60, Bonn.
Piso, I., Tecar, T. 2013, *Une inscription funéraire des environs de Napoca*, Tibiscum SN 3, forthcoming.
Piso, I., Cupcea, G. 2014, forthcoming, *Ein centurio regionarius aus der legio X Fretensis in Dakien.*
Pitts, L.F., St. Joseph, J.K. 1986 - Inchtuthil. The Roman Legionary Fortress Excavations 1952-65, Britannia Monograph Series 6, London.
Pohlsander, H.A. 1996, The Emperor Constantine[2], London/New York.
Pollard, N. 2000, Soldier, Cities and Civilians in Roman Syria, Michigan.
Popescu, M. 2004, La religion dans l'armée Romaine de Dacie, Bucureşti.
Probst, P. 2009, *Der optio in der römischen Armee. Untersuchung zur Stellung innerhalb der Rangordnung sowie zur Funktion im militärischen und administrativen Bereich. Ein Vorbericht*, Limes 20 Léon, 1111-1115.
Radin, M. 1915, *The Promotion of Centurions in Caesar's Army*, CJ 10/7, 300-311.
Rainbird, J.S. 1986, The Fire Stations of Imperial Rome, PBSR 54, 147-169.
Rankov, N.B. 1990, *Singulares legati legionis: A Problem in the Interpretation of the Ti. Claudius Maximus Inscription from Philippi*, ZPE 80, 165-175.
Rankov, N.B. 1990a, *Frumentarii, the castra peregrina and the Provincial officia*, ZPE 80, 176-182.
Rankov, N.B. 1994, *Die Beneficiarier in den literarischen und papyrologischen Texten*, Schallmayer 1994, 219-232.
Rankov, N.B. 1999, *The governor's men: the officium consularis in provincial administration*, The Roman Army as a Community (ed. I. Haynes, A.K. Goldsworthy), JRA SS 34, Portsmouth/Rhode Island,15-34.
Rea, J.R. 1980, *Ordinatus*, ZPE 38, 217-219.
Reddé, M. 1995, *La Rangordnung des marins*, Le Bohec 1995, 151-154.
Reuter, M. 2012, Legio *XXX Ulpia Victrix*. Ihre Geschichte, ihre Soldaten, ihre Denkmäler, Xantener Berichte 23, Xanten.
Richardot, P. 1995, *Hiérarchie militaire et organisation légionnaire chez Végèce*, Le Bohec 1995, Paris, 405-427.
Richier, O. 2004, *Centuriones ad Rhenum*. Les centurions légionnaires des armées Romaines du Rhin, Paris.
Rostovtzeff, M. 1957, The Social and Economic History of the Roman Empire[2], Oxford.
Roth, J.P. 1994, *The Size and Organization of the Roman Imperial Legion*, Historia 43, 346-362.
Roth, J.P. 1999, The Logistics of the Roman Army at War (264 BC-AD 235), Leiden/Boston/Köln.
Roxan, Margaret M. 1995, *The Hierarchy of the Auxilia. Promotion Prospects in the Auxilia and work done in the last twenty years*, Le Bohec 1995, 139-146.
Sablayrolles, R. 1996, *Libertinus miles*. Les cohortes des *vigiles*, Rome.
Saddington, D.B. 1982, The Development of the Roman Auxiliary Forces from Caesar to Vespasian (49 BC-AD 79), Harare.
Saddington, D.B. 1990, *The Origin and Nature of the German and British Fleets*, Britannia 21, 223-232.
Saddington, D.B. 1995, *Problems in military ranks and military personnel in Josephus*, Le Bohec 1995, 53-56.
Saddington, D.B. 1996, *Early Imperial praefecti castrorum*, Historia 45, 244-252.
Saddington, D.B. 2007, *Classes. The Evolution of Roman Imperial Fleets*, Erdkamp 2007, 201-217.
Sage, M.M. 2008, The Republican Roman Army. A Sourcebook, New York/London.
Southern, P. 1989 - *The Numeri of the Roman Imperial Army*, Britannia 20, 81-140.
Sander, E. 1952, *Triarius ordo*, RheinMus 95, 79-96.
Sander, E. 1954, *Zur Rangordnung des römischen Heeres: Die gradus ex caliga*, Historia 3/1, 87-105.
Sander, E. 1957, *Zur Rangordnung des römischen Heeres: Die Flotten*, Historia 6/3, 347-367.
Sander, E. 1959, *Zur Rangordnung des römischen Heeres: Der Duplicarius*, Historia 8/2, 239-247.
Sarnowski, T. 1993, *Nova Ordinatio im Römischen Heer Des 3. Jh.s und eine neue Primus Pilus-Weihung aus Novae in Niedermoesien*, ZPE 95, 197-204.
Sarnowski, T. 1993a, *Primi ordines et centuriones legionis I Italicae und eine Dedikation an Septimius Severus aus Novae in Niedermoesien*, ZPE 95, 205-219.
Sarnowski, T. 1996, *Promotio ex nova ordinatione eines künftigen Centurio. Zur Inschrift aus Novae in ZPE 95, 1993*, ZPE 111, 289-290.
Šašel, J. 1972, *Zur Rekrutierung der Prätorianer*, Historia 21/3, 474-480.
Šašel Kos, M. 1995, *The beneficiarii consularis at Praetorium Latobicorum*, Römische Inschriften – Neufunde, Neulesungen und Neuinterpretationen. Festschrift für H. Lieb (ed. R. Frei Stolba, M.A. Speidel), 149-170.
Saxer, R. 1967, Untersuchungen zu den Vexillationen des römischen Kaiserheeres von Augustus bis Diocletian, Epigraphische Studien 1, Köln/Graz.
Schallmayer, E. 1990 (ed.), Der römische Weihebezirk von Osterburken I. Corpus der griechischen und lateinischen Beneficiarier-Inschriften des Römischen Reiches, Stuttgart = CBFIR.
Schallmayer, E. 1994 (ed.), Der römische Weihebezirk von Osterburken II. Kolloquium 1990 und paläobotanische-osteologische Untersuchungen, Stuttgart.
Schallmayer, E. 1994a, *Die Beneficiarer in Obergermanien*, Schallmayer 1994, 161-192.
Schmidt, J. 1886, *Die Rangklasse der Primipilaren*, Hermes 21/4, 590-596.

Schmidt Heidenreich, C. 2000 – *Hiérarchie et structure des unités palmyréniennes de l'armée Romanie de Dacie*, Army and Urban Developement in the Danubian Provinces of the Roman Empire, Alba Iulia, 293-308.

Sherk, R.K. 1974, *Roman Geographical Exploration and Military Maps*, ANRW II.1, 534-562.

Sinnigen, W.G. 1961, *The Roman Secret Service*, CJ 57/2, 65-72.

Sinnigen, W.G. 1962, *The Origins of the "Frumentarii"*, MAAR 27, 211, 213-224.

Sinnigen, W.G. 1967, *Tirones and Supernumerarii*, CPh 62/2, 108-112.

Smith, R.E. 1972, *The Army Reforms of Septimius Severus*, Historia 21/3, 481-500.

Smith, R.E. 1979, *Dux, praepositus*, ZPE 36, 263-278.

Southern, Pat 1989, *The numeri of the Roman Imperial Army*, Britannia 20, 81-140.

Southern, Pat 2001, The Roman Empire from Severus to Constantine, London/New York.

Southern, Pat 2006, The Roman Army. A Social and Institutional History, Santa Barbara/Denver/Oxford.

Spaul, J.E.H. 1994, *Ala²*. The Auxiliary Cavalry Units of the Pre-Diocletianic Imperial Roman Army, Farnborough.

Spaul, J.E.H. 2000, *Cohors²*. The evidence fo and a short history of the auxiliary infantry units of the Imperial Roman Army, BAR Int.S. 841, London.

Speidel, M.A. 1992, *Roman Army Pay Scales*, JRS 82, 87-106.

Speidel, M.A. 1993, *Miles ex cohorte. Zur Bedeutung der mit ex Eingeleiteten Truppenangaben auf Soldatenischriften*, ZPE 95, 190-196.

Speidel, M.A. 1995, *Rang und Sold im römischen Heer und die Bezahlung der vigiles*, Le Bohec 1995, 299-310.

Speidel, M.A. 2000, *Sold und Wirtschaftlage der römischen Soldaten*, Alföldy, Dobson, Eck 2000, Stuttgart.

Speidel, M.A. 2001, *Specialization and Promotion in the Roman Imperial Army*, Administration, Prosopography and Appointment Policies in the Roman Empire (ed. L. De Blois), Amsterdam, 50-61.

Speidel, M.A. 2007, *Albata decursio – ein kaiserliches Siegesmanöver. Zu einer neuen Inschrift aus Ankara*, ZPE 162, 263-270.

Speidel M.A. 2009, *Augustus' militärische Neuordnung und ihr Beitrag zum Erfolg des Imperium Romanum,* Heer und Herrschaft im römischen Reich der hohen Kaiserzeit, MAVORS XVI, 19-52.

Speidel, M.P. 1970 - *The Captor of Decebalus. A new Inscription from Philippi*, JRS 60, 142-153.

Speidel, M.P. 1973 - *The Pay of the Auxilia*, JRS 63, 141-147.

Speidel, M.P. 1974, *Stablesiani. The Raising of new Cavalry units during the Crisis of the Roman Empire*, Chiron 4, 541-546.

Speidel, M.P. 1976 - *Eagle-Bearer and Trumpeter. The Eagle Standard and Trumpets of the Legions Illustrated by Three Tombstones Recently Found at Byzantion*, BJb 176, 123-163

Speidel, M.P. 1978, Guards of the Roman Armies, Antiquitas 28, Bonn.

Speidel, M.P. 1978a, *Felix legio vestra. A Building Inscription from Romula-Malva*, ZPE 30, 119-122.

Speidel, M.P., Dimitrova-Milčeva, Alexandra 1978, *The Cult of the Genii in the Roman Army and a New Military Deity*, ANRW II.16.2, 1542-1555.

Speidel, M.P. 1980 - *A Spanish Cavalry Decurion in the Time of Caesar and Augustus*, AEA 53, 211-213.

Speidel, M.P. 1981, *Princeps as Title for ad-hoc Commanders*, Britannia 12, 7-13.

Speidel, M.P. 1982 - *Legionary Cohorts in Muretania. The Role of Legionary Cohorts in the Structure of Expeditionary Armies*, ANRW II, 10.2, 850-860.

Speidel, M.P. 1982a, *The Career of a Legionary*, TAPhA 112, 209-214.

Speidel, M.P. 1983, *The Centurions' Titles*, Epigraphische Studien 13, 43-61.

Speidel, M.P. 1984, *Regionarii in Lower Moesia*, ZPE 57, 185-188.

Speidel, M.P. 1984a, *Germani corporis custodes*, Germania 62/1, 31-45.

Speidel, M.P. 1986, *Centurions and Horsemen of legio II Traiana*, Aegyptus 66, 163-168.

Speidel, M.P. 1986a, *Rangzeiche für Zenturionen un die große Weihinschrift aus dem Mainzer Legionslager*, JRGZM 33/1, 321-330.

Speidel, M.P. 1986b, *Maxentius and his "equites singulares" in the Battle at the Milvian Bridge*, ClA 5/2, 253-262.

Speidel, M.P. 1990, *The Names of Legionary Centuriae*, Arctos 24, 135-137.

Speidel, M.P. 1992, *The Framework of an Imperial Legion*, The 5th Annual Caerleon Lecture,5-47.

Speidel, M.P. 1992a - *Centurions Promoted from Beneficiarii?*, ZPE 91, 229-232.

Speidel, M.P. 1992b, *The Weaponskeeper, fisci curator, and the Ownership of Weapons in the Roman Army*, MAVORS VIII, 131-136.

Speidel, M.P. 1992c, The Career of a *strator* and *summus curator*, MAVORS VIII, 137-139.

Speidel, M.P. 1994, Riding for Caesar. The Roman Emperors' Horse Guards, London.

Speidel, M.P. 1994a, *The Tribunes Choice in the Promotion of Centurions*, ZPE 100, 469-470.

Speidel, M.P. 1994b - *A Guardsman as Officer of Irregulars*, ZPE 103, 215-216.

Speidel, M.P. 1994c, *Legionary Horsemen on Campaigns*, SJb 47, 36-39.

Speidel, M.P. 2002, *Becoming a Centurion in Africa. Brave Deeds and the Support of the Troops as Promotion Criteria*, Roman Army Studies II, MAVORS VIII, 124-128.

Speidel, M.P. 2005, *Centurial Signs and the Battle Order of the Legions*, ZPE 154, 286-292.

Sperber, D. 1969, *The Centurion as a Tax-Collector*, Latomus 28, 186-188.

Steidl, B. 2005, *Die Station der beneficiarii consularis in Obernburg am Main. Vorbericht über die Ausgrabungen 2000/2002*, Germania 65/1, 67-94.

Stein, A. 1927, Der römische Ritterstand, München.

Stauner, K. 2004, Das offizielle Schriftwesen des römischen Heeres von Augustus bis Gallienus (27 v.Chr.-268 n.Chr.), Bonn.

Stoll, O. 1995, *Die Fahnenwache in der römischen Armee*, ZPE 108, 107-118.
Stoll, O. 1997, *Die Benefiziarier – Rangordnung und Funktion. Einige Bemerkungen zur neuen Forschung*, Laverna 8, 93-112.
Stoll, O. 2001, *Ordinatus architectus. Römische Militärarchitekten und ihre Bedeutung für den Technologietransfer*, MAVORS XIII, 300-368.
Stoll, O. 2003, *"Medicus centurio" (PSI 1063). Ein sanitätsoffizier mit taktischem Kommando? Probleme, Hypothesen, Lösungen*, JRGZM 50/1, 329-354.
Stoll, O. 2007, *The Religions of the Armies*, Erdkamp 2007, 451-476.
Strobel, K. 1984, Untersuchungen zu den Dakerkriegen Trajans. Studien zur Geschichte des mittleren und unteren Donauraumes in der Hohen Kaiserzeit, Antiquitas 33, Bonn.
Strobel, K. 1987, *Bemerkungen zur Laufbahn des Ti. Claudius Vitalis*, Tyche 2, 203-209.
Strobel, K. 1988, *Ein weiteres Zeugnis zur Rangordnung im römischen Legionscenturionat der Kaiserzeit*, EA 12, 43-45.
Strobel, K. 1995, *Rangordnung und Papyrologie*, Le Bohec 1995, 93-112.
Strobel, K. 2007, *Strategy and Army Structure between Septimius Severus and Constantine the Great*, Erdkamp 2007, 267-285.
Sumner G.V. 1970, *The Legion and the Centuriate Organization*, JRS 60, 67-78.
Syme, R. 1932, *Die Zahl der praefecti castrorum im Heere des Varrus*, Germania 16, 109-111.
Syme, R. 1939, *The Roman Revolution*, Oxford.
Thomas, C. 2004, *Claudius and the Roman Army Reforms*, Historia 53/4, 424-452.
Traverso, M. 2000, *Il centurionato nelle legioni romane: la legio II Augusta*, Athenaeum 88, 219-252.
Tully, G.D. 1998, *The stratarches of legio VI Ferrata and the Employment of Camp Prefects as Vexillation Commanders*, ZPE 128, 226-232.
Țentea, O., Matei-Popescu, F. 2004, *Alae et cohortes Daciae et Moesiae*, ActaMN 39-40/I, 259-296.
Ubl, H. 2001, *Gedanken zu einem Benefiziarierabzeichen aus Lauriacum*, Carinthia Romana und die Römische Welt. Festschrift für G. Piccottini, 379-390.
Verboven, K. 2007, *Good for Business. The Roman Army and the Emergence of a Business Class in the Northwestern Provinces of the Roman Empire (1st Century BCE – 3rd Century CE)*, Impact of Empire 6 (ed. De Blois, Lo Cascio), Leiden/Boston, 295-314.
Watson, G.R. 1958, *The Pay of the Roman Army. The Republic*, Historia 7/1, 113-120.
Watson, G.R. 1959, *The Pay of the Roman Army. The Auxiliary Forces*, Historia 8/3, 372-378.
Watson, G.R. 1965 - *Immunis librarius*, Britain and Rome. Essays presented to Eric Birley on his 60th Birthday, Kendal, 45-55.
Watson, G.R. 1969, The Roman Soldier, Bristol.
Watson, G.R. 1974, *Documentation in the Roman Army*, ANRW II.1, 493-507.
Webster, G. 1969 (1988), The Roman Imperial Army. Of the First and Second Centuries A.D., London.
Webster, G. 1986, *Standards and Standard-Bearers in the Alae*, BJb 186, 105-115.
Wegeleben, Th. 1913, Die Rangordnung der Römischen Centurionen, Berlin.
Wesch-Klein, Gabriele 1998, Soziale Aspekte des römischen Heerwesens in der Kaiserzeit, HABES 28, Stuttgart.
Wesch-Klein, Gabriele 2000, *Commeatus id est tempus, quo ire, redire quis possit. Zur Gewährung von Urlaub im römischen Heer*, Alföldy, Dobson, Eck 2000, 459-472.
Wiegels, R. 2013, *Zu einer Statuette des Sol im Museum August Kestner in Hannover*, forthcoming.
Willmans, J.C. 1987, Zur Rangordnung der römischen Millitärärzte während der mittleren Kaiserzeit, ZPE 69, 177-189.
Wolff, Catherine 2004, *Primus pilus ou primipilus?*, REMA 1, 5-22.
Young, G.K. 1997, *The Customs-Officer at the Nabataean Port of Leuke Kome (Periplus Maris Erythraei 19)*, ZPE 119, 266-268.
Zwicky, H. 1944, Zur Verwendung des Militärs in der Verwaltung des römischen Kaiserzeit, Zürich.

www.ingramcontent.com/pod-product-compliance
Lightning Source LLC
LaVergne TN
LVHW071216210926
848482LV00015B/825